A Practical Guide Book for Engineers
in Apparel Manufacturing

Industrial Engineer's Digest

Learn, Practice and Improve Factory Performance

Prasanta Sarkar

Postgraduate in Fashion Technology (NIFT, Delhi)
B. Tech in Textile Technology (Calcutta University)

Copyright© 2021 by Prasanta Sarkar
All rights reserved.

Cover design: Sam Smith and Vardan

This book is published by Online Clothing Study under Online Clothing Study (OCS) education series.

No part of this book may be reproduced in any form by any means without the permission of the author.

For information contact
Online Clothing Study
Email: ocs@onlineclothingstudy.com
Website: www.onlineclothingstudy.com

ISBN-10:
ISBN-13:

First print edition: 2021

Disclaimer:
Due care and diligence have been taken while editing and printing the book. Neither the Author, nor the Printer of the book holds any responsibility for any mistake that may have crept in inadvertently. Online Clothing Study, the publisher will be free from any liability for damages and losses of any nature arising from or related to the content.

This book is sold subject to the condition that it shall not by way of trade or otherwise, be lent, resold, hired out, circulated, and no reproduction in any form, in whole or in part (except for brief quotations in critical articles or reviews) may be made without written permission of the publishers.

MRP: ₹850.00

To
the Global Fashion Industry Professionals

More from Prasanta Sarkar

1. Garment Manufacturing: Processes, Practices and Technology
2. Garment Maker's KPI: Why Measure and How to Measure
3. Production Planning and Control in Apparel Manufacturing
4. Industrial Engineering: Guide to Job Interview Preparation
5. The Beginner's Guide to Leggings Manufacturing

Table of Contents

Acknowledgement 1
Introduction 3
Abbreviations 5

Part-I: Industrial Engineering Basics

Know Your Responsibilities 9
Duties and Responsibilities of an IE Department 14
Top 5 Activities an IE Does in a Garment Factory 18
Tools and Equipment used by IEs 20
Application of IE Concepts in Finishing Section 21
Involvement of IE in the Cutting Department 23
Time Study of Garment Operations 25
Method Study of Garment Operations 29
What is the Standard Time (SAM)? 31
Standard Minute Value (SMV): Definition, Calculation and Use 33
What is Bundle Allowance in SAM? 38
Takt Time and Its Application 39
The Concept of Operator's Performance Rating 41
How to Use Performance Rating for Calculating Basic Time? 43
Operation Breakdown 45
Sequencing of Stitching Operations 48
Making an Operation Bulletin of a Garment 52
The Secret of Making Operation Bulletin for Beginners 57
Line Layout 59
What is the Balancing Loss? 61
Sewing Process Flow Chart of a Garment 64
Sewing Machines Used in Making Polo Shirts 66
Sewing Machines Used in Making T-Shirts 68
Line Balancing 69
Different Types of Garment Production Systems 71
Benefits of Single-Piece Flow Production System 77
Different Types of Industrial Sewing Machines and Their Use 79
Different Type of Sewing Line Layouts 83
Different Kind of Material Handling Systems 89
Grading of Sewing Operators 92
Skill Matrix of Sewing Operators 95
How to Develop Skill Matrix? (using an Excel Template) 97
Garment Production Cost: Actual Cost Vs Cost Per SAM 102
The Way Factories Calculate Production Cost 104
On-Standard Efficiency and Overall Efficiency 106
Major Factors Affecting Factory Efficiency 109
Difference between Productivity and Efficiency 111
Pitch Time and Pitch Diagram 113
What is Standard Hour Earned (SAH)? 116
Estimation of Cutting SAM 118

v

Role of the Garment SAM in Production Planning 122
Incentive Systems – An Introduction 124
An Incentive Scheme for Sewing Operators 127
Attendance Bonus for Workers 132
Non-Productive Time (NPT) 134
Why Measure Non-Productive Time of a Line? 139
What is Needle Downtime? 141
How to Calculate Needle Downtime? 142
The Concept of Idle Time 144
Therbligs and 18 Motions Name with Symbols 145
What is TMU in Time and Motion Study? 147
What is a Learning Curve? 148

Part-II: Data Capturing, Calculations and Reports

Production Reports Made by IE Department 153
Necessary Production Reports for a Garment Factory 155
Tips to Make Daily Production Report (DPR) Quickly? 159
Hourly Production Report 162
Calculating Labor Cost Using Work Measurement 165
Measuring Productivity in a Garment Factory 167
Difference between SAM Productivity and Production Productivity 172
Monthly Efficiency Report 174
Calculate Line Efficiency for Multiple Styles 179
Method of Calculating Cost per Minute of a Sewing Line 181
Use of Display Boards on the Sewing Floor 183
KPIs for Garment Manufacturers (Key Performance Matrices) 185
KPI Dashboard for the Sewing Factories 188
Calculation of Man to Machine Ratio (MMR) 191
How to Calculate Man to Machine Ratio for a Double Shift Plant? 193
Calculate Number of Days Needed to Complete an Order 194
Calculate WIP in Cutting, Sewing and Finishing Section 196
7 Ways to Reduce WIP from Bottleneck Operations 198
Why Measure WIP in a Garment Factory? 201
How to Make WIP Report in Garment Production? 202
Reporting and Data Analysis of Overtime Work 205
Order Completion Report 212
Useful Formula for IEs and Production Team 214

Part-III: How to Guides

How to Calculate Garment SAM? 223
How to Estimate Sewing Time by Using Machine RPM? 228
How to Reduce Standard Time (SAM) of a Style? 229
How is the Sewing Time Calculated (Machine Time in SAM)? 230
Measure SMV Improvement Percentage 232
How to Estimate Daily Garment Production? 234

How to Calculate Production Target and Worker's Bonus in the Initial Days of Production Start? 236
How to Calculate Target Production per Hour from Cycle Time 239
How to Balance a Traditional Sewing line? 240
How to do Line balancing using Operator Skill History? 243
How to Calculate Stitching Machine SPI? 249
How to Calculate an Operator Efficiency? 251
How to Show Line Efficiency when there is no Loading to a Line? 255
How to Calculate Overall Line Efficiency? 257
How to Get Maximum Efficiency in Shorter Run Orders? 262
Importance of Accurate Data in Reporting 264
How to Calculate Machine Requirement for Sewing Operations? 265
How to Calculate Cost of Manufacturing Apparel Products? 268
How to Calculate Number of Sewing Lines Needed for an Order? 271
How to Reduce Line Setting Time? 273
How to Control Apparel Production Cost? 275
How to Calculate Sewing Floor Capacity? 277
How to Calculate Production Capacity of a Factory? 279
How to Calculate Hourly Production Target? 284
How to Calculate Sewing Operator Requirement? 287
How to Estimate Machine Requirement for a Factory? 289
How to Calculate Helper Requirement for a Sewing Line? 291
How to Calculate Manpower Requirement of Finishing Section? 292
Relationship Between Machines, Working Days and Output 293
How to Calculate CM Produced by a Line in Dollar? 295
How to Calculate Garment Production Cost from SMV and Monthly Salary? 298
How to Do Production Scheduling Without SAM? 300
How to Improve Line Efficiency in a Piece Rate Factory? 302
How to Plan Daily Line Output from Garment SAM? 305
How to Establish Thread Consumption Ratio 307
How to Calculate Thread Consumption of a Garment? 309
How to Justify the Piece Rate of a Garment? 311
How to Encourage Workers in a Garment Factory? 314
Sewing Operator Recruiting Test Procedure 317

Part-IV: Improve Factory Performance

Skill Upgradation of Sewing Machine Operators 321
Can an Operator Hit 100% Efficiency? 324
Following up the Hourly Production Target 326
20 Ways to Improve Shop Floor Productivity 327
20 Tips to Increase Shop Floor Performance 335
9 Ways to Increase Sewing Operator Efficiency 337
9 Things You Can Do to Control Production in a Factory 339
Why Measure Stitching Line Efficiency? 342
How to Improve Finishing Room Efficiency? 343
Selection of Improvement Projects and Project Implementation Method 344
Need Ideas for Improvement Projects – Check This List 347

Part-V: Advanced Reading

Some Factories Do Things Differently 351
Which payment system is better- Piece Rate System or Salary based? 353
Managing Individual Performance Bonus 355
Line Efficiency: A Magical Measure for Garment Manufacturers 358
Work Study and Industrial Engineering Terms and Definitions 361
7 Must Read Books for Industrial Engineers 364
Sewing Machine Ratio for a Shirt Factory 366
Relationship Between Line Efficiency and Labor Costing 368
How had I Learned Industrial Engineering (and How You Can Learn Too)? 369
Difference Between Basic Time and Standard Time 372
Pros and Cons of Setting up Garment SAM Higher than the Standard Time 374
Key Qualities Needed to Be a Successful IE Manager 376
Identify the Reasons for Low Line Efficiency 378
Characteristics of a Good Workstation Layout 380
Operator Training and Deployment in Apparel Industry - A Systemic View 382
Recording Operator's Work Start/Stop Time and Its Importance 386
Should We Include Helpers in Line Efficiency Calculation? 389
Who Should Count Operator's Production? 392
Review Yearly Performance by using 5 Trend Charts 393
What is More Important- a Skilled Operator or a Sewing Machine? 395
Change Factory Floor Layout and Give it a New Look 397
Real-time System for Production Tracking System in Apparel Manufacturing 399
Software to Establish Garment Standard Time 403
About Prasanta Sarkar 405

Acknowledgement

The ideas, knowledge, and information shared in this book are learned from working with industrial engineers in garment factories, by reading books and articles of many authors, and from open sources. I learned engineering from many IEs, teachers, and seniors, and shared my learnings in this book in the form of articles and short notes. What I have done is packaged useful information, guides, and ideas for apparel production students, industrial engineers, and garment production professionals in apparel manufacturing.

Thanks to all these knowledge sources.

I am thankful to all my teachers and instructors. Thanks to Paul Collyer, Dr. Rajesh Bheda, Dr. Prabir Jana, Manoj Tiwari, Manoj Kumar Singh, Santosh Kumar Subudhi, Crishantha Perera, Vineet Saini, Durga Charan Das, Ram Chandra Das, Laszlo Szabo, Keerthi Abe, Madhaiyan C, Rajesh Kumar, Achyutananda Nayak, and many others for enriching my knowledge on this subject.

Thanks to all the learners who keep me asking questions on the industrial engineering topics the apparel manufacturing. Many active learners and industrial engineers from various countries inspired me in writing. Appreciation from readers of my earlier work helped to keep me active and writing hundreds of articles on the IE topics on the apparel manufacturing industry. Thanks to all OCS readers.

Special thanks to my family members for allowing me to work late hours for writing articles and editing this book after office hours.

Thanks to Vardan Singh and Sam for designing the book cover. Thanks to the Pothi team for taking care of the printing and distribution of this book.

Introduction

This book is written for you if you want to learn the industrial engineering basics, about the necessary tools for engineers and activities done by industrial engineers. This book is for you if you want to work as an industrial engineer in a garment factory.

By learning industrial engineers subject, you can bring changes and bring improvement in the factory where you are working and where you will be working.

An engineering degree is not necessary to improve a factory's productivity and reducing the manufacturing cost. What is required is the right attitude.

If you allow yourself to learn industrial engineering tools, you can learn most of them in one month. Then you can practice these IE tools and IE activities in the next 3 months. After that, you are ready for serving the apparel manufacturing industry. You can make things better in a garment factory.

You need to find ways of doing things in a better way - which in turn can bring a huge improvement. If you can improve line efficiency by 1% each week, monthly efficiency improvement will be 4%.

In a factory, to bring measurable improvement you need to fight against the odds, resistance from the line supervisor, and non-acceptance of new things and new concepts. To fight against these odds, you need to be strong within yourself through being more knowledgeable, logical, analytical, and proactive.

This book will enrich your knowledge. The how-to guide part will increase your confidence in finding solutions and answers to the odd questions at the workplace.

I started writing articles and blog posts in 2011. If you have visited my blog and got a chance to explore my work, you already know that I have covered many subjects in my writings including apparel production, industrial engineering, apparel merchandising, apparel quality control, garment production, production planning, technology solutions for the apparel industry and MIS. The majority of articles that I have written are on industrial engineering subjects. This book is a collection of articles written on the OCS blog. All the topics and articles are edited and fine-tuned in this book.

This book is not like a standard textbook written for students, with chapters and sections. You will not get overwhelmed with the contents of this book. Topics are written in the form of short articles, how-to guides, answer to questions received from OCS readers. You can start reading any topic from the book selecting it from the content page.

You will find many tricks and tips on production management, how-to guides for performing industrial engineering tasks. IE topics and problems faced by newcomers are explained to the point.

In our professional life, every day we need some information, and guides to accomplish our task. We need to increase our productivity in our existing work. At the same time, we take new responsibilities. We need to manage a team. In our busy days, we do not get time for reading books and explore more on the subject areas.

Throughout the book, I have included images, tables, and screenshots. The numbering of these tables and images are done within the article. So, refer to the table and figure number within the article where more than one item is found.

Abbreviations

Abbreviations and short form of common terms used in this book and their full forms.

DHU – Defects per Hundred Units
DPR – Daily production report
HOD – Head of the Department
IE – Industrial Engineering / Industrial Engineer
IED – Industrial Engineering Department
KPI – Key Performance Indicators
M/c – Machine
MIS – Management Information System
MMR – Man to machine ratio
MTM – Methods time measurement
NPT – Non-productive time
OB – Operation bulletin
OCS – Online Clothing Study
OT – Overtime (hours)
PBS – Progressive bundle system
PMTS – Predetermined motion time system
PPC – Production planning and control
RFID – Radio frequency identification
RMG – Readymade garment
SAH – Standard allowed hour (Standard hours earned)
SAM – Standard allowed minute
SMV – Standard minute value
SOP – Standard operating procedure
SPI – Stitches per inch
TMU – Time measuring unit
UBT – Underbed trimmer
UPS – Unit production system
WIP – Work in progress

Abbreviations of Machine Names

Full forms of machine names used in this book.

SNLS – Single needle lock stitch
SNCS – Single Needle Chain Stitch
DNLS – Double Needle Lock Stitch
4TOL – 4 Thread Over Lock
FL – Flatlock (3TFL – 3 threads flatlock)
BH – Buttonholing machine
BS – Button Stitching machine
SNLS (EC) – Single needle lock stitch with Edge cutter
SNLS(UBT) – Single Lock Stitch with underbed trimmer
FOA – Feed off the Arm
IRON – Ironing/Pressing table
Manual – Manual works (like marking, folding, pressing)

PART-I

Industrial Engineering Basics

Know Your Responsibilities

Whether it is by choice or by chance, we are in the apparel manufacturing sector. And we are part of the fashion business supply chain.

As you are reading this book, I am assuming that you are an industrial engineer or a work study officer in a garment factory or you want to be part of the garment manufacturing industry. This book is especially written for industrial engineers and other garment industry professionals who like to learn IE things and apparel production management.

For explaining the overall responsibilities of an industrial engineer, I will start with applications of industrial engineering concepts in the garment manufacturing. I will cover the roles and responsibilities of the IE department. Then I will show you the list of common industrial engineering tools used by industrial engineers.

Application of industrial engineering concepts

Industrial engineering concepts are used in garment manufacturing to fulfill the following purposes -

- Monitoring production floor and having better control over the production floor
- Improving processes and improving method of working to increase the factory's overall performance and standardized garment manufacturing processes
- Overall application of the industrial engineering concepts can be explained better by describing common tasks of the IE department.

Industrial engineer's activity list

Common tasks of the IE department are as the following but not limited to these.

- Work measurement of sewing operations, cutting room jobs, and finishing jobs.
- Establishing standard time (SAM) for sewing operations and manual operations.
- Style analysis and conducting research and development (R&D) of the garment samples.

- Improving working methods of doing operations and any kind of activities.
- Improving workstation layout.
- Production planning and factory capacity determination.
- Work aids selection for the new styles and development.
- Labor cost estimation.
- Measuring workers performance.
- Setting line production target of the sewing lines and chasing production from line supervisors and operators
- Product study and understanding its construction
- Conducting Time study
- Time and Motion analysis of the operations
- Method Study (observation of movements of an operation)
- Preparing operation breakdown
- Preparation of OB (Operation bulletin) – estimation of manpower requirement, machine requirement and requirement of the folders, guides and sewing machine attachment.
- Preparing machine layout, workstation layout design and implementation of the same.
- Production line setup when loading a new order to a line.
- Production estimation of a style based on resources availability.
- Work Sampling
- WIP Control
- Line Balancing – levelling the workload across the production line.
- Capacity study – conducting capacity study of sewing operators in sewing operations after style loading.
- Developing and maintaining operators' skill matrix
- Designing a feasible incentive schemes for workers and calculating incentives earned by workers.
- Estimating thread average for production orders
- Sometimes IEs are involved in operator training

How are IE concepts used by apparel manufacturers?

Factories those are using IE techniques mostly have a complete IE set up (department). An IE department consists of an IE manager (in-charge), industrial engineers and junior engineers. The strength of IE team widely varies based on the maturity level of the department and on the focus of application of industrial engineering. Without having enough team members, an IE department cannot work effectively.

Normally, industrial engineers are utilized in the following ways. Factories apply all or some of the above listed activities to -

- assist line supervisors by preparing resource requirement plan (machines, equipment, and manpower), line setting and line balancing etc.
- assist production managers in target planning and production planning,
- help merchandiser and marketing personnel by providing labor cost and production lead time,
- help HR department by providing operator performance level, and help in operator recruitment
- prepare MIS reports and show the factory management production status on daily basis and alert management if their attention is needed.
- set up standards operating procedures for new tasks.
- Following up the new process/ updated process required for ever-changing fashion products

More than the regular jobs, IEs are also responsible for thinking of continuous process improvement. Initiation of new projects and implementation of the project that has been undertaken, showing the improvement opportunity within the factory to a management team. Engineers are involved in performance improvement tasks of the cutting department and finishing department.

Companies those have limited manpower in IE team, they involve engineers only for routine jobs like, making operation bulletin preparation, thread consumption, preparing daily production reports etc.

Roles of IEs in production management

I had a discussion with a couple of industrial engineers working in different garment factories. From the discussion, it was understood that engineers have a bigger role in managing the garment manufacturing processes. They are involved in production management in a garment factory.

In the daily schedule, they do constant monitoring of all the production processes. Some of the common managerial roles of an IE include.

1. Recruitment of sewing operators through the standard test method (operator skill test).
2. Evaluation of existing production target and benchmarking a new production target.
3. Extend work measurement in other production processes like cutting section and finishing section.
4. Constant control of direct labor cost.
5. Data analysis and reviewing monthly trends of each production KPIs.
6. Preparing daily production reports, monthly improvement reports with graphical presentation.
7. Time study and updating time study database including videos on regular basis.
8. Research on motion study and implementation of the better system to eliminate excess motions.
9. Research on new machinery and the latest systems that meet companies long term goal and preparing cost-benefit analysis and ROI for the same.
10. Initiative on lean manufacturing, learning of lean tools and looking for the scope of implementation.
11. Research and development for samples prior to production start.

There are additional activities for which engineers are responsible for. If any project is being implemented in a factory related to production, industrial engineers lead the project.

No doubt that industrial engineering department is adding a lot of values to the company. Without having the industrial engineering team, one would not know how things are going on the shop floor, what corrective measures need to be taken on time, effective utilization of manpower and machines.

Industrial engineering concept needs to incorporate at the time of plant set up instead of changing and modifying things later after factory starts making garments.

A written list of job responsibilities of a department is always beneficial for managing and monitoring a department from the top level. This also helps employees to be focused on their core responsibilities.

Duties and Responsibilities of an IE Department

This is a detailed list of duties and responsibilities of the industrial engineering department in a garment factory. You can use this list and implement it in your factory right now. You may need to modify some points based on your organization structure and IE team's strength. Job responsibilities are listed according to IE's different positions. This list is shared by Madhaiyan C, a long time OCS reader.

Duties and responsibilities of the Industrial Engineering Head
Following are the responsibilities of the IE Head in his company.
- Analyzing the CMT and Productivity details for all new development styles based on garment sketch or samples and provide CMT to merchandising department.
- After order confirmation the evaluated garment is re-analyzed for better and economical methods to reduce the negotiated CMT in order to stay in safe margin. It means reduction of SAM by construction simplification related to seam and method without affecting the actual appearance of sample.
- Releasing style OB, Line Layout and thread consumption chart for line(s) and all factories (having multiple factories) based on scheduled plan.
- Department wise Manpower requirement rationalization based on target strength ratio of 1:0.8
- Co-ordination with planning department for allocation of similar styles, capacity plan, line plan and factory monthly plan based on the history of line and factory efficiency
- Co-ordination with planning department for improving the pre-production activities.
- Co-ordination with Factory manager for cut plan and finishing plan based on delivery priority.
- Co-ordination with HR manager for training progress (fresh training and multi-skill training).
- Co-ordination with HR manager to reward outstanding operators based on Skill grade, efficiency, and attendance (once in 15 days).
- Factory audit to be conducted to ensure the IED OB and Layout are followed in line.

- Monitoring and Auditing IE and Jr.IE activities in all the units
- Factory Production Incentive report monitoring and approval
- Working on new projects if any and suggesting advanced techniques in garmenting

Duties and Responsibilities of Industrial Engineer

Industrial engineers are responsible for the following activities.

- Operation Bulletin (OB) preparation to be done for three stages, Initial Proto Sample and FIT, approved sample, and actual OB for layout.
- Layout preparation for every style
- Every month machinery requirement plan to be prepared and arrange to in house necessary machine in advance by coordinating with Factory Manager and Planning head
- Arrange to in-house folders and attachment well in advance based on scheduled production plan
- Thread consumption chart preparation and co-ordinate with merchant team for thread purchase.
- Skill Matrix updating once in three month
- Updating the SAM Database based on method improvement.
- Preparing and updating sampling status based on production plan (PP/size set samples to be ready with approval one week before the style loading date).
- Preparation of factory wise incentive reports based on respective department production and efficiency.
- Monitoring weekly performance report of individuals and conduct meeting to improve their performance.

Duties and Responsibilities of Junior IEs (Sewing Floor)

Junior Industrial Engineers of sewing floor are responsible for the following activities.

- Three days before the style starts in line, layout should be discussed with Factory Manager and Production Manager and get their signature to implement
- If any changes in the layout is done, it must be updated and change accordingly with the approval of IED – Head

- Requesting machines, folders, and attachment requirement well in advance based on scheduled production plan
- Conducting Method study and educate to follow the best method to operators
- Conducting capacity study of individuals and fix their target after 3 days from the style starts
- Follow-up study (operators who are all not achieving the Target/Capacity)
- Bundle system follow-up
- Bottle neck operation – identify and eradicate
- Fixing new tailors grade based on their performance and criteria.
- Developing multi-skill operators based on their individual interest and maintaining skill matrix record.
- List out low efficiency operator and monitor to improve their efficiency
- Measuring line lost hours due to various reasons in sewing floor
- Monitoring and Line balancing – Based on capacity and WIP in the line
- Helping supervisors in line balancing during high operator absenteeism

Duties and Responsibilities of Junior IEs (Cutting and Finishing Department)

Industrial Engineers of Cutting and Finishing departments are responsible for the following activities.

- Cutting Operation Bulletin preparation
- Cutting capacity checking and manpower allocation depending on the style
- Finishing Operation Bulletin preparation
- Finishing capacity study and manpower allocation depending on style and priority of the order delivery
- Training program of new trainee tailors (15 days basic - pedal control, paper, and cloth exercise) to be monitored
- After successful basic training, training exam to be conducted and trainees those scores more than 50 marks out of 100 to be transferred to production line

- Transferred trainee tailors are involved in production floor for a month period and monitoring their daily production and efficiency. During this period, head counting, and target are not fixed to them.
- After one month, efficiency curve chart to be prepared and their average efficiency more than 35%, those operators are transferred to line and designated as Tailor grade E (entry level).

Duties and Responsibilities of Data Entry Operator

Data entry operator is also considered as a part of the IE department. His activities include as following.

- Preparation of daily individual efficiency report
- Preparation of daily line and factory efficiency report
- Preparation of daily line and factory lost time report
- Preparation of style costing report
- Preparation of MIS Meeting report
- Preparation of incentive working report

In case your company has multiple factories, and you have industrial engineers with the similar positions, you can follow the above lists for job responsibilities.

Top 5 Activities an IE Does in a Garment Factory

We know industrial engineers (IE) do a lot of work in a garment factory. They take care of work-study activities, process improvement, MIS system and report preparation, and production management etc.

If you are asked what are the top activities IEs do in your factory, what will be your answer?

Daily activities one needs to perform depend on the job responsibilities and job profile of the engineer in the department. If you are an IE manager, your job will be more on managing team, coordinating with other departments and data analysis area. But if you are a junior IE or a work-study officer, most possibly you will spend your time on Time study, capacity study, making operation bulletin, making the daily production report, thread consumption calculation etc.

Depending on the maturity level of IE department in a company and company's primary focus in improvement and growth, factory involves its IE team in different activities. So, IE's jobs will be different from one company to another.

In fact, there are many activities performed by the IE department. But you know what the most important tasks are that you do in your factory.

I used to visit garment factories often and work with industrial engineers. As an outsider (visitor) followings are top 5 activities I have seen IE department does on daily basis

1. Report analysis (KPI data)
2. Controlling production cost
3. Line balancing
4. Method improvement
5. Production floor management

I had conducted a survey asking the following question (top 5 activities done by IEs). I had received many responses on this survey. See the survey result in the following table.

Question: If you are an industrial engineer, list your top tasks. If you are not an IE, but you are aware of IE's common activities as per your experience and knowledge what are 5 tasks industrial engineers do?

Industrial Engineer's Digest

Survey Information:

Here I am sharing the few entries received from industrial engineers. From the lists provided by some engineers; we can assess the common activities performed by industrial engineers in the garment factory. This would surely help engineers for setting up their top priority in the work.

Job Profile	Activity #1	Activity # 2	Activity # 3	Activity # 4	Activity # 5	Other Activities
Senior IE	Daily/Weekly/Monthly report Analysis	CMT Preparation	Time Study, Method Study	Operation Breakdown	Budget Preparation	Line Balancing, Production Floor Control, R&D for New styles
Senior IE	Daily Production report	Production Floor management	Replacement of the absenteeism	Line balance	Follow the trainees	Export Planning follow UP
Senior IE	Preparation New Sample	Make video proses	Planning Layout and Mapping	Thread Consumption	Line Balance and Improvement	KPI
Junior IE	Production report	Time study	Line balancing	Line capacity	Efficiency	Line target
Senior IE	Prepare Operation Breakdown	Thread Consumption	Line Balancing	CMT given	Balancing	Costing
Senior IE	costing production	Balance line				
Junior IE	Report analysis	Controlling production cost	Line balancing	Method improvement	Production floor management	Innovation
IE Manager	Operator skill management & line balancing	Fallow KPI process	Cost control	Style risk analysing & developments	Avoid loss time & improve productivity	
Senior IE	Line Layout	Capacity study/Follow up	Bottleneck identification and elimination	Line balancing/ Manpower balancing	Throughput time & Changeover time calculation	Process standardization
Senior IE	Line balancing	Make paper layout	Report analysis (KPI data)	Cost minimization	Method improvement	Time & motion study
IE Manager	Manpower Balancing	Daily Forecast Plan from Cutting to packing	Ontime line setting	Efficiency Achievement	Method improvement and cost controlling	Plan to performance
Junior IE	Line Balancing	Identify Bottleneck Area & Try to Remove the Problem	Capacity Study	Making Operation Bulletin	WIP Report	

19

Tools and Equipment used by IEs

In your day to day job as an industrial engineers (IE), you will use various type of tools and equipment. What all equipment and tools to use, depend on the job responsibilities of an IE in a company. Name of the common tools and equipment are listed below. Purposes of the tools are also mentioned here.

- **Stopwatch (Digital or analog):** Measuring observed time at the time of Time Study. Also used for capacity study.
- **Measuring Tape:** Measuring the length of seams and measuring distances for workstation layout.
- **Digital Camera:** Capturing videos for various operations that help in motion analysis of operations.
- **Tripod for the camera:** Used as a camera stand.
- **Time Study board:** Required during time study to hold the Time Study format.
- **Calculator:** For data calculation and report making.
- **Tachometer:** to measure the speed of the motor of the sewing machine. This equipment is used to find machine rpm.
- **Data capturing and process analysis formats:** For example- Time Study format, Motion analysis format etc. Data capturing and analysis to bring improvement.
- **Various documents:** Documents are used to assist production and other processes with information, methodology or layout. Formats are like Operation Bulletin, Line Layout, Pitch Diagram, Hourly report format etc.
- **Computer:** Data analysis, report making, emailing, video analysis of operations etc.
- **PMTS system** – If the factory owns a PMTS software
- **Real-time system** – for data capturing and report data
- **Planning tool** - if you are responsible for planning activities.

The technology of time study, data analysis and data capturing are changing. You may need to handle a new tool in the coming days. But if you have the above tools, you are ready for doing your job.

Application of IE Concepts in Finishing Section

Question: *I am working as Asst. Officer (IE) at ABC Fashion Ltd (name changed). I just want to know the application of IE in the finishing department. I mean what we can do as IE in the finishing department. It will be very helpful if you reply me back. Thanks ...asked by Hasan.*

You are an industrial engineer. You already know about industrial engineering tools and techniques used for managing and improving production. Most IEs practice industrial engineering concepts only in the sewing floor. Similar tools can be applied in finishing section also.

In the following, I have shared few points that can be done in finishing department without further brainstorming and skill upgrading.

1. **Setting of Standard Time (SAM) for finishing jobs (activities)-** Like sewing operations, estimate SAM for finishing jobs. Use Time Study method or synthetic data (PMTS) for determining SAM of the finishing tasks. Prepare operation sheet for finishing section also. By establishing SAM of finishing job, you can plan finishing department's production accurately. You can estimate finishing cost and manpower requirement.
2. **Study existing finishing room workstations layout** - Observe the scope of improvement in finishing room workstation layout. You can apply theory of motion economy for re-designing finishing room workstation layout.
3. **Suggest a better material handling method to finishing room workers** - There is always some scope for betterment.
4. **Work on method improvement** - Observe all finishing tasks. Like garment ironing, spot removing, garment checking, thread cutting, folding, and packing activities. Study the method of working different activities. If you see workers do not follow the same methods for doing the same task, you can standardize working method for finishing room activities. Spend time on improving a method. By improving method of doing a task, you can reduce cycle time of a task.
5. **Maintain WIP in between two workstations** – capture finishing production data and calculate WIP between the

finishing processes. You can work on levelling of workload in each process.
6. **Train checkers and other workers** – enhance the skills of workers by training on their task.

Other things you can do - making hourly production report, capacity study, preparing efficiency report and labor productivity analysis for finishing department.

Start with one activity. Once you are involved in working in the finishing section, you will find many ideas in improving finishing room productivity by applying industrial engineering techniques.

Involvement of IE in the Cutting Department

Question: What all activities industrial engineer can do in the cutting department? - This question was asked an OCS reader.

In the garment industry, industrial engineers mainly focus on the sewing department. IEs perform various activities in the sewing floor and work on improving production processes. IE skills and knowledge can be applied in the other departments too, like cutting department and finishing department in a garment factory.

Most of the garment factories employ industrial engineers for sewing department and they are supposed to be responsible for sewing floor activities only. They can initiate improving process in cutting department by introducing industrial engineering concepts, IE tools and techniques.

When it comes activities of cutting department, IEs can be involved in the following activities.

1. **Time Study for cutting room activities and setting Standard Time** – Cutting room activities like fabric spreading, cutting, re-cutting, ply numbering and bundling can be standardized. The standard time for each activity can be measured.
2. **Making standard operating procedures (SOPs) for all cutting activities** – You might know the importance of SOPs for processes and activities to produce the desired output. IE department can involve themselves in preparing SOPs for cutting department.
3. **Measuring KPIs** – An IE can measure cutting room KPIs. Like marker efficiency%, fabric utilization percentage, cutting quality, and machine and manpower utilization.
4. **Designing cutting room layout** - Cutting room layout can be re-designed if needed to improve workflow from one section to another section.
5. **Suggesting good material handling equipment** – Good material handling equipment can improve cutting room performance; improve quality of cut parts etc. You can find suitable equipment for cutting department too. For example, you can work on improving manual layering method, cutting process, and bundling and numbering method.

6. **Finding a solution for cutting room issues**- There is always a scope of improvement in cutting room performance. As an IE you can start practicing problem-solving to solve day to day cutting room issues.
7. **Studying and analyzing manpower requirement in cutting room** – Scientific calculation of manpower requirement in the cutting room is neglected. You can introduce it and reduce excess manpower from the cutting room.

These activities can be included under IE's responsibility. To get benefit out of IE's involvement in cutting section, they need to concentrate fully for cutting department only. You need to build a separate IE team who would perform these tasks.

Industrial Engineer's Digest

Time Study of Garment Operations

Definition of Time Study

Time study is a method of measuring work for recording the times of performing a certain specific task or its elements carried out under specified conditions. An operator does the same operation (task) throughout the day. Time study helps to define how much time is necessary for an operator to carry out the task at a defined rate of performance.

Time Study Tools

To do time study you need to arrange the following tools
- Stopwatch (Digital stopwatch)
- Time study format
- One pen or pencil
- Time Study board

How to Conduct Time Study?

An operation cycle consists of material handling, positioning, and aligning parts, sewing garment parts, trimming threads, and tying and untying a bundle. So, in the time study format, divide whole task into various elements according to the motion sequences of the operation.

I will show you steps for conducting time study. In this study, we will take and operation 'collar run stitch'. The seamline is shown in the right side image. Let us break the operation cycle into the following 5 elements.

Figure 1: Line diagram of a collar

- pick up the collar panels to sew first seam,
- turn collar to sew second seam,
- turn collar to sew third seam
- check work and dispose and
- waiting for next pieces.

I will show you filled time study format for collar run stitch. I did this study during my internship in a garment factory.

Step 1: Preparation
- Arrange stationery like time study format, stopwatch (digital one) and pencil
- Select one operation for Time study
- Tell the operator that you are going to measure the time he/she takes to do the job.
- Observe the operation carefully and breakdown each operation into elements.
- Fill the basic information in the time study format. Like machine category, guide or attachment used.

Step 2: Record Time

Now measure the time taken for completing each element of an operation cycle by the operator. Time should be captured in seconds. Similarly, capture element timing for consecutive 10 operation cycles. During data capturing only note down reading (see following Table-1) of the stopwatch and later calculate element time. If you found any abnormal time in any elements, record time during time study and later discard that reading. Or you can capture time for one more cycle.

Abnormal reading may be captured due to bobbin change, thread break, power cut or quality issues. Ignore those reading from the calculation of standard time.

Time Study Sheet												
Sheet No. 02				Machine: SNLS								
Operation: Collar run Stitch				Attachment/Guide: N/A					Operation image			
Fabric: Poplin				UBT/Non-UBT: UBT m/c								
Style Description: Full Sleeve Shirt				Style No.: XYZ								

	Operation Element Breakdown									Foreign elements		
	Pick up & stitch 1st line		Stitch 2nd line		Stitch 3rd line		Check and dispose		Waiting			Description
Operation Cycles	T	R	T	R	T	R	T	R	T	R	T	R	
1	5	5	4	9	3	12	9	21	6	27	1		
2	4	31	7	38	3	41	8	49	0	49	2		
3	6	55	3	58	2	60	11	71	5	76	3		
4	6	82	4	86	3	89	6	95	0	95	4		
5	7	102	5	107	3	110	7	117	0	117	5		
6	6	123	4	127	4	131	8	139	3	142	6		
7	6	148	5	153	4	157	7	164	6	170	7		
8	5	175	5	180	2	182	8	190	0	190	8		
9	6	196	4	200	5	205	8	213	5	218	9		
10	6	224	4	228	32	260	8	268	1	269	10	260	

Table-1: Time Study sheet (data recording part)

Step 3: Calculation of Basic Time

From the Reading (R) calculate time taken for each element for all five cycles just by deducting previous Reading from elemental reading. Sum up times of five cycles for each element. Note, if you discard any reading then in that case number of cycles will be four. Calculate average element times. This average time is called observed time. (in the following Table-2 it is noted as average time). To get the better accuracy in cycle time, you can record up to 20 operation cycles.

Time Study Sheet												
Sheet No. 02				Machine: SNLS								
Operation: Collar run Stitch				Attachment/Guide: N/A					Operation image			
Fabric: Poplin				UBT/Non-UBT: UBT m/c								
Style Description: Full Sleeve Shirt				Style No.: XYZ								
	Operation Element Breakdown										Foreign elements	
	Pick up & stitch 1st line		Stitch 2nd line		Stitch 3rd line		Check and dispose		Waiting			Description
Operation Cycles	T	R	T	R	T	R	T	R	T	R	T	R
1	5	5	4	9	3	12	9	21	6	27	1	
2	4	31	7	38	3	41	8	49	0	49	2	
3	6	55	3	58	2	60	11	71	5	76	3	
4	6	82	4	86	3	89	6	95	0	95	4	
5	7	102	5	107	3	110	7	117	0	117	5	
6	6	123	4	127	4	131	8	139	3	142	6	
7	6	148	5	153	4	157	7	164	6	170	7	
8	5	175	5	180	2	182	8	190	0	190	8	
9	6	196	4	200	5	205	8	213	5	218	9	
10	6	224	4	228	32	260	8	268	1	269	10	260
	Summary											
Total Time	57		45		61		80		26			
Abnormal Time	0		0		32		0		26			
Number of Cycles	10		10		9		10		10		Allowances	
Avg. Time	5.7		4.5		3.22		8		0		Machine Allowances	
Rating	100%		100%		100%		100%		100%		SNLS : 9%	
Normal Time	5.7		4.5		3.22		8		0		DNLS: 14%	
Allowances(%)	20%		29%		29%		20%		0		OL: 3T= 7%, 4T=9%	
Standard Time (Elements)	6.84		5.805		4.16		9.6		0		5T=11%	
Standard Time (in Sec.)	26.40											
Standard Time (in Min)	0.44										Fatigue: 20%	

Note: T=Time, R= Clock Reading. Allowances added here are for example.
Time Study format source: NIFT, New Delhi.

Table-2: Time Study sheet with calculation

Step 4: Calculation of Standard Time

To convert the average observed time (cycle time) into normal time, you need to multiply the basic time by the operator performance rating. Here, for calculation purpose, operator rating is considered as 100%.

Add allowances to normal time (machine allowances, fatigue, and personal needs etc.) Add machine allowance only to those elements where machine is running and fatigue and personal needs to all elements (approx. 20%). Machine allowances are shown in the time study sheet. (I have used these allowance percentage as I have seen factory IEs use this allowances.)

The calculated time (Normal time + Allowance time) is standard time for each element in seconds. Sum up all elemental time and convert seconds into minutes. This is the standard time (in minute) of the selected operation. This is called as Standard Allowed Minute (SAM).

Objectives of Time Study

The main objective of time study is to establish standard time of a garment operation. In this section, you read how the standard time of an operation is established. Other common objectives include-

- Operation cycle time is measured through time study method.
- For capacity study of each operation, we use time study tool
- We calculate potential capacity of each operator in selected operations.
- Manpower requirement and machine requirement estimation.

Method Study of Garment Operations

Method study is more of a systematic approach to job design than a set of techniques. It is defined as the systematic recording and critical examination of existing and proposed methods of doing work, as a means of developing and applying easier and more effective methods and reducing costs (Work study definition by ILO). The method study involves systematically following six steps

1. Selection of work to be studied:

Most operations consist of many discrete jobs or activities. The first stage is to select those jobs to be studied that will give the best returns for the time spent. For example, activities with the best scopes for improvement, those causing delays or bottlenecks or those resulting in high costs.

2. Recording of all relevant facts of the current method:

Method study uses formal techniques to record the sequence of activities, the time relationship between different tasks, the movement of materials, the movement of staff. There are many techniques used in method study.

3. Critical examination of those facts:

This is the most important stage in the method study. It is used to critically examine the current method by seeking answers to questions:
- The purpose of each element
- The place
- The sequence
- The person
- The means

4. Development of the most practical, economical, and effective method:

This stage is used to develop a new and a better method of executing the task, by considering the results of the critical examination. The new method is developed by a combination of entirely eliminating some activities, combining some parts, changing the sequence of some activities and by simplifying the content of others.

5. Installation of a new method:

This step involves project managing the changes and ensuring that everybody involved understands the changes involved. In other words, they understand the new method, which is doing what, the differences compared to the old method and crucially the reason for the changes. Training is an important part of this stage particularly if the new method involves radical changes. Providing modified equipment, components and layouts may also be involved.

6. Maintenance of new method and periodic checking:

Monitoring of how effective the new method is and how personnel has adapted is very important. One aspect that is sometimes overlooked is to check what effect the new method has on other activities. For instance, it may be that whilst the new method is successful in eliminating a bottleneck in a specific area, the bottleneck has moved elsewhere in the process. By periodic checking the new method and its effects, management can ensure that overall efficiency is improving rather than deteriorating.

Here I have only described the method involved in method study in brief. To practice method study (time and motion study), you need to read more on this subject, and you need to understand it in detail. Read the book *'Introduction to work study by ILO' by George Kanawaty*.

What is the Standard Time (SAM)?

What is SAM (Standard Minute)?

The meaning of standard allowed minute (SAM) defined in the book Introduction to Work Study, ILO. Standard minute is defined as the following.

SAM is the time value arrived at for a task based on the average rate of output which qualified workers will naturally achieve without overexertion provided that they know and adhere to the specified method and provided that they are motivated to apply themselves to their work.

Here **SAM** stands for **S**tandard **A**llowed **M**inute.

In the garment manufacturing, specifically in garment production, SAM is used for measuring work. In a factory, work study department (IE department) measures and calculates SAM for sewing operations using standard calculation method. SAM is used many ways directly and indirectly. Some of those are -

- Factory as well as individual operator performance is measured by using SAM data
- Labor cost and associated cost ratios are measured using standard time (SAM)
- Operators' payroll and incentive amount is calculated based on earned SAM
- Latest production scheduling systems like line balancing system and performance measuring systems use SAM as one of the primary parameters

An organized garment manufacturing company cannot think of managing and controlling shop floor without using SAM of the garment and individual operations. Even apparel buyers do negotiation of garment manufacturing cost based on garment SAM. To derive most appropriate SAM value of a garment, factories use PMTS based systems. Standard Time is also derived from time study method.

Standard Minutes (SAM) of a Few Basic Garment Products

It is not advisable to estimate the garment SAM without seeing and/or analyzing the garment.

To estimate the SAM of a garment you must analyze the garment carefully and check different factors that affect the garment SAM.

SAM of a product varies according to the work content or simply according to number of operations, length of seams, fabric types, stitching accuracy needed, working method, sewing technology to be used.

Though we know these facts, sometimes beginners need approximate SAM values for basic products, like Tee Shirt, Formal shirt, Formal trouser or jacket.

An estimated SAM helps in capacity planning of the factory, calculating the requirement of machinery and even helps to estimate CM (cut and make) costing of a garment.

Standard minutes (SAM) of a few basic products is listed here with its SAM range according to work content variation. In actual cases, garment SAM may go outside of the given limit.

Table: Average SAM for few garment products

Serial No.	Product	SAM (Average)	SAM range
1	Round neck T-shirt	8	6 – 12
2	Polo shirt	15	10 – 20
3	Formal full sleeve shirt	21	17 – 27
4	Formal trouser	35	
5	Sweatshirt (Hooded)	45	35 – 55
6	Jacket (suit)	101	70 – 135
7	Women Blouse	18	15 – 45
8	Bra	18	16 – 30
9	Coat (for chefs and doctors)	21	

The garment product SAM shown in the above list is for reference purpose only.

For your SAM library, you can also develop similar list of various products. You can develop this list for the products that you are making in your factory. Also include other products that you are not making currently.

Standard Minute Value (SMV): Definition, Calculation and Use

There is no more secret in Standard Minutes Value (SMV) calculation. Now, SMV is a common language between fashion brands and manufacturers for discussions on labor cost, time required to make one garment and shop floor capacity.

Most of the garment manufacturing companies calculate SMV foe garments they make. Industrial engineers take responsibility for measuring standard time for the given garment samples.

Let me define SMV, the most critical factor, widely used and much talked about term in the garment industry.

Definition of Standard Minute Value

Standard Minute *is the time value arrived at for a task based on the average rate of output which qualified workers will naturally achieve without over exertion provided that they know and adhere to the specified method and provided that they are motivated to apply themselves to their work.* - This definition is written in the ILO work study book.

Operation cycle time or average cycle time, taken by an operator completing an operation is not the Standard Minute value. SMV is the given time that should be taken to complete operation when an operator works at 100% performance level.

SMV of the same operation will be different if the working condition is changed – like operator using a different machine, the operator sews a bigger component, an operator using attachment and work-aids when sewing a garment.

Most importantly, movements and motions involved in performing a task effect the SMV of an operation.

SMV calculation method

You can calculate garment SMV using SMV calculation software or through time study method. Standard minute value calculation method using Time Study is explained in another article with an example.

First, you measure operation wise SMV. Then SMV of all operations in style are combined to calculate total SMV of a garment.

In time study, you capture the observed time of the selected operation. You rate operator performance in the work. You use standard allowances for different types of machines, bundle allowance and contingency allowances.

Steps for calculating SMV

1. Record observed time.

Following the Time Study procedure, you conduct Time Study for a selected operator and on the selected operation.

2. Convert observed time into Basic time using this formula.

The observed time for the same operation done by two different operators might be different. Our objective is to establish the standard for the selected operation. So, for the levelling of the observed time, it is multiplied by the rating factor. The new time is called as basic time.

Basic Minute = (Observed time in minutes x Performance Rating)

3. Know the applicable allowances.

When an operator will be working there would be chances of a machine breakdown, they may need to change the needle, replace empty bobbin, change thread spool, may need of threading needle. There will be personal needs and fatigue allowance. For all these allowances, some amount of time will be consumed from the operator's standard time. To estimate more practical SMV, allowances are added to the basic time.

Normally, in progressive bundle system, garment components reached to an operator in bundle form. For every bundle, the operator spends a few minutes in opening the bundle and tying it. So, you need to add bundle allowance as well where applicable.

4. Calculate Standard minute

Add machine allowance, bundle allowance and other allowances to the basic time to calculate standard minute.

Standard Minute Value = (Basic minute + Bundle allowances + machine allowance + personal fatigue allowances).

Normally, when SMV is measured through time study method, the operation cycle is broken down into operation elements and observed time is captured element wise. Machine allowances are added only to machine elements.

Also, note that bundle allowance is distributed across the pieces in a bundle. If bundle allowance is 60 seconds for a 10 pieces bundle, 6 seconds will be added to the basic time as bundle allowance.

Example: Let's say you are making a blouse. You observed an operation and recorded its cycle time. Following the above steps let's derive SMV of an operation.

- Observe Time = 2 Minutes
- Rating of the operator = 90%
- Basic Time = (2 Minutes x 90%) = 1.8 Minutes
- Machine allowance = 10%
- Other Allowance = 10%
- Total allowance to add = 10% + 10% = 20%
- Allowance in minutes = 1.8 x 20% = 0.36 Minutes

Calculated SMV = (1.8 Minutes + 0.36 Minutes) = 2.16 minutes

In the operation bulletin, after measuring operation SMV, hourly and daily production capacity is calculated.

Read articles on the capacity calculation, machine requirement calculation topics.

The Second method

The second method of establishing SMV is using PMTS software. For this method, you need to know different motion codes and you need to study real-life operator hand and body movement. You need to learn how to operate software for calculating sewing operation SMV. Once you learn using a PMTS system, you can measure SMV for any kind of operations whether it is cutting, pressing, sewing, folding etc. (Essential thing you must have the software).

Icon	Code	Code Description	Time (Seconds)	
		SEW TWO PARTS TOGETHER	Machine	Manual
	MG2T	Match and Get 2 Parts Together		1.95
	FOOT	Put Part(s) to Machine Foot		1.30
	MS1A	Sew to hold	0.90	
	AM2P	Align and Match 2 Parts		1.52
	S9LA	Sew 9cm, Low Guidance and Stop > 1cm	1.35	
	AM2P	Align and Match 2 Parts		1.52
	S9LA	Sew 9cm, Low Guidance and Stop > 1cm	1.35	
	TCUT	Cut Thread		1.65
	APSH			

A presentation of standard time calculation using motion codes | Image source: YouTube/GSDCost

For your reference, establishing SMV using PMTS software, steps are shown in this article: *How to calculate accurate standard time?*

SMV calculation Software

If you want accurate SMV of garment (and other sewn product), you need a software to calculate the SMV. Most of the SMV calculation software uses motion codes and predetermined times (MTM2). Few of the well-known SMV calculation software included - GSDCost, SewEasy, ProSMV, and TimeSSD.

Application of SMV Data

Some of the applications of garment SMV.
- SMV is used in factory capacity planning
- SMV is the main data in the operation bulletin. Based on the operation SAM machine requirement and manpower requirement is calculated. Further based on the machine requirement machine layout and initial line balancing is done.
- SMV is used for labor costing for garment production
- SMV is used for performance benchmarking
- SMV is used for efficiency calculation
- SMV is used in Payroll calculation in piece rate payment. Earning is equal to (Garment produced quantity **x** operation SMV **x** Rate per minute)

- When performance incentive is provided to the operators, SMV is essential for calculating individual operator efficiency
- Preparing operators' performance-based skill matrix

Who is responsible for calculating garment SMV in a garment factory?

Normally, in a garment factory, industrial engineers or work-study officers calculate the garment SMV. Further, they do production studies during bulk production on the floor and validate garment SMV. SMV is measured in two stages – first at the time when buyer request garment costing by sending a garment design or referring a garment sample. The second time, SMV is measured prior to loading style into the production line. Further, the SMV is updated by engineers after production study.

Definition of different terms used in SMV calculation process

Rating factor: Rating is a subjective comparison of any condition or activity to a benchmark, based upon our experience. While the mechanics of time study record the time a task did take, applying a rating will determine the time a task should take.

Cycle Time: It the time taken by an operator at the work. Cycle time is measured from the time stamp of pick-up of the garment component to the next pick-up time stamp of an operation cycle. The starting point of an operation cycle can be chosen by the engineer. Normally the cycle time value is higher than the standard time.

Observed time: Here observed time is the same as the cycle time. The time recorded as it was, and operation cycle time calculated.

What is Bundle Allowance in SAM?

You may already know that to determine Standard Time (SAM) of an operation through Time Study method, various allowances are added to the basic time. Bundle allowance is one such allowance. Do you know what bundle allowance is?

In Progressive Bundle Production System (PBS), number of garments are bundled together. These bundles are fed to the sewing line. Sewing operators need to open (untie) the bundle before she starts sewing garment pieces. After completing stitching of one bundle, she needs to tie up the bundle. Then she stacks the bundle or slides it to the following operator. For this activity, the operator spends a certain amount of time for each bundle handling. It may be a fraction of minutes. The time required for bundle tying and untying is considered bundle allowance.

Bundle Allowance Time per piece will depend on the size of the bundle. A bundle may contain 5 pieces or 20 pieces of garments (Sometimes even more). For both bundles, same time will be required to untie and tie the bundle. When we calculate standard time, it is done for single pieces. Bundle allowance will be less as bundle size increases.

Example: During establishing the operation SAM, a standard bundle size is considered. Normally, 10 pieces a bundle. In that case, if the bundle allowance 1 minute for 10 pieces bundle, bundle allowance per piece would be one tenth of 1 minute (6 seconds).

Industrial Engineer's Digest

Takt Time and Its Application

What is Takt Time?

Takt time is the allowable times to produce one product at the rate of customers' demand. This is NOT the same as the cycle time, which is the normal time to complete an operation on a product (which should be less than or equal to TAKT time).

Takt time is the calculated pace of production based on the average speed at which the customer is buying a product or service. The formula is net available time to produce per time period divided by customer demand per time period.

Let's say net available time is 4500 minutes per shift (10 operators' total man-minutes) and the customer's demand 500 pieces per shift. In this case, the Takt time will be equal to (4500÷500) = 9 minutes per piece. This means in every 9 minutes; you need to produce one garment.

Important things to be noted that,
- Takt time cannot be measured with a stopwatch.
- Takt time is not the time it takes to perform a task.
- Takt time is only reduced or increased by changes in the production demand or net available time to work.

How takt time is used in garment production?

As the definition says, it is the demand of customer or demand of the following process. To setup an assembly line, takt time is considered as a base to determine work content to be given to each operator in the line.

Takt time is a very important tool for a Lean Line where single-piece flow system is used.

For example, demand from the production line is 60 pieces per hour. In an hour, you have 3600 seconds. So, takt time for the line will be 60 seconds/piece (3600 seconds/60 pieces).

According to this target and garment SAM (suppose men's full sleeve shirt), you need to determine how many operators is needed for setting up an assembly line.

Let's assume SAM of a shirt style is 20 minutes (1200 seconds). In one minute, each operator has only 60 seconds. So, to produce a piece in

60 seconds, the total number of operators required 1200/60 = 20 Nos. (Considering that each operator works at 100% efficiency.)

You can distribute all the operations involved in making a complete shirt within 20 operators. All operations will have a different work content (SAM). So, to equalize work content each operator will get work of about 60 seconds of work content. For this line set-up, few operators need to do multiple operations with low work content. And in some operations, you may need to assign more than one operator.

I hope now you understand how the Takt Time can be used in a garment factory.

The Concept of Operator's Performance Rating

Definition of Performance Rating:

Rating is a subjective comparison of any condition or activity to a benchmark, based upon our experience. While the mechanics of time study record the time a task did take, applying a rating will determine the time a task should take by a 100% rated performer.

Let's consider walking is an activity. You are observing 3 people walking on the way – through observation, you can assess who walking faster, who is slower and who is walking speed is average. This is a kind of rating of walking speed.

When we establish standard time of a garment operation, we use performance rating factor for standardizing standard allowed time for a given task.

Basic Time= (Cycle time **x** Performance Rating)

What is 100% performance or Normal Performance?

The concept of 100% performance is a critical element of time study and performance measures. Normal performance is the rate of output which qualified workers will achieve without over-exertion over the working day shifts provided they know and adhere to the specified method and provided they are motivated to apply themselves to the work. This performance is denoted as 100% on standard rating and performance scales.

A slower performance rate, which will produce fewer pieces per hour, is recorded as a percentage below 100%. A faster performance rate that produces more pieces per hour is recorded as greater than 100%.

Characteristic of 100% Performance or Normal operator

- Fluid motions without hesitation
- No false starts or duplications
- Consistent, coordinated, effective rhythm
- No wasted actions or work
- Attention centered on the task

How to get accurate rating?

To improve accuracy in rating an operator, observer must -
- Has knowledge of the operation and the specified method or standard operating procedures for that task.
- Concentrates on operator motions
- Is alert to fumbles, hesitations, and other lost motions- these are seldom or absent in 100% performance.
- Eliminates or ignores interruption or events, not in the operator's control.
- Avoids a corrupting bias when observing fast and slow operators in succession
- Knows that increasing the number of cycles observed increases accuracy

In the next article, I have covered a topic how to apply performance rating in calculating the basic time of a sewing operation. In another article, you will get the answer why rating is multiplied to the observed time to calculate basic time of a task.

Reference: tc2.com

How to Use Performance Rating for Calculating Basic Time?

In the article *'Calculation of SAM using Time Study'*, I have applied operator performance rating to calculate basic time. In that example, operator performance rating is 80% and observed cycle time is 0.60 minutes. Basic time is calculated as (0.60 x 80%) = 0.48 minutes.

Many OCS readers asked me the question, why rating factor is multiplied to cycle time instead of dividing when calculating the basic time using performance rating. If you also have the similar question in mind, then read the following two questions and read my answer to those questions. This article explains why we need to multiply cycle time by rating factor to calculate basic time.

Question#1:
In your article, you indicate that if an operator has an 80% rating that you multiply the cycle time by the rating (which reduces the cycle time for the operation). Wouldn't you actually divide the cycle time by the rating to increase the cycle time for the operation before you add the other allowances? ... asked by Rick.

Question#2:
Actually, I am confused why basic time getting lower when the performance rating is lower in your article, whereas it should be another way, which is basic time should be more when the rating resulting in lower. So instead of multiplying you need to divide the basic time from rating. Can you please explain this as I do not understand this? ... asked by Kumar.

Answer:
I replied to Rick and I had a discussion with him to clarify the confusion over whether to multiply or divide the rating% to cycle time to calculate basic time. Later, I had just sent this discussion to Kumar. He understood this completely and accepted my explanation. I hope you will also understand it once you read the following conversation.

Prasanta: Hi Rick, Rating factor is used to normalize the cycle time. 100% rated operator's time considered as normal time (basic time). So, if an operator is working at 80%, she is actually doing her work at a slower pace compared to a 100% performing operator. That's why she is taking a longer time than the 100% performer. Secondly, the normal time to do the job is less than the time taken by her (80% rated operator). That is why the rating factor is multiplied to the cycle time instead of dividing. I hope you understand my point.

Rick: Thank you for responding. Yes, I believe I understand what you are saying, however, I still believe that you need to divide the standard cycle time by the operator rating.

Let's take an example... If the cycle time of an operation is 10 seconds and the operator is rated as 80% then if as you described above, he/she is working slower than the normal cycle time, so his/her overall time must be LONGER than the normal cycle time, correct? 10 seconds divided by .80 equals 12.5 seconds (longer). If we were to multiply the cycle time by the rating, we would get a shorter cycle time (10 seconds times .80 equals 8 seconds). I am very new to all of this so I want to be sure that I'm getting this correct. Thanks in advance for the help.

Prasanta: Rick, you are not following the purpose of application of Rating. The purpose is to normalize the cycle time or making equivalent to 100% rated operator. Here, cycle time is observed time. The purpose to not increasing or lowering cycle time. The cycle time is the observed time taken by an operator at her current rate (pace) of working. Our aim is to establish standard time of that operation from her cycle time that would be the same to the cycle time of a 100% rated operator.

Just ask this question - in your example at 80% operator is completing a task in 10 seconds. How much time she would take if worked at 100%? Result must be lower than 10 seconds. That is why you need to multiply rating to cycle time to calculate normal time.

I hope my explanation helps you understood the application of rating in SAM calculation. See few examples of basic time calculation in the following table.

	Operation	Cycle Time (Seconds)	Performance Rating	Basic Time (Seconds)
1	Run stitch collar	20	80%	16 (20x0.8)
2	Crease collar	24	110%	26.4 (24x1.10)
3	Top stitch collar	30	100%	30 (30 x 1.0)
4	Attach front placket	40	90%	36 (40 x 0.90)

Operation Breakdown

Each garment is constructed following number of sewing jobs and non-sewing jobs. Jobs and activities involved in making a garment is known as operations. E.g., shoulder attach, and bottom hem are example of operations. In mass garment manufacturing, prior to making the garments, the garment sample is studied carefully and list of required operations for making the approved design are made. The operations are written in a sequence of actual process flow to be followed while making the garment in the shop floor.

What is operation breakdown?

The method of preparing operations' list in a sequence is called as operation breakdown.

The sheet of listed operations of a style is also known as operation breakdown. An operation breakdown includes information like

- Sewing and non-sewing operations
- Name of the machines to use for doing the specific operations
- Estimated time to do each operation for one unit

One sample operation breakdown of Polo shirt is shown in the following Figure-1.

The garment operation bulletin is also called as operation breakdown. An operation bulletin includes more information than the operation breakdown. In another article, the method of making operation bulletin is explained.

Use of operation breakdown

Some of the primary use of an operation breakdown includes -

- The operation breakdown of a garment is made to understand garment construction - like stitch class and seam types used in making the sample garment.
- To make an operation bulletin, we first prepare garment operation breakdown.

Operation Breakdown (Polo Shirt)

Sequence No.	Operations	Estimated Time	Machine type
1	Fuse Placket	0.13	Fusing M/c
2	Hem Front Placket	0.55	SNLS
3	Bottom Hem	0.64	Flatlock
4	Size label to main label	0.22	SNLS
5	Main label to yoke	0.28	SNLS
6	Attach placket to front	0.83	SNLS
7	Cut placket	0.40	Manual
8	Front plkt edge stitch	0.51	SNLS
9	Sew Box at plkt	0.68	SNLS
10	Yoke attach to back	0.46	DNLS
11	Shoulder overlock	0.56	5T O/L
12	Topstitch Shoulder	0.34	SNLS
13	Tack piping at collar ends and Attach piping to collar	0.77	SNLS
14	Turn Placket ends inside out and Finish Piping	1.00	SNLS
15	Cuff O/L	0.45	5T O/L
16	Cuff Topstitch	0.59	SNLS
17	Sleeve attach	0.89	5T O/L
18	Sleeve topstitch	0.60	SNLS
19	Slits Making	1.40	SNLS
20	Side seam O/L & bottom plkt O/L	0.86	4T O/L
21	Tack at sleeve ends	0.30	SNLS
22	Button Hole	0.29	BH
23	Button Attaching	0.30	BS
	Total SAM	13.05	

Fig-1: Sample operation breakdown sheet of a Polo Shirt

- To make the list of sewing machines and equipment required for the style. Machine type is identified from seam type and stitch class.
- To set up a line for a new style, operation breakdown is followed by line supervisors and engineers.

- While estimating thread consumption per garment for a given style, operation breakdown is referred for operation list and machine types.
- Operation breakdown is used as a guide for tailors for constructing a complete sample.
- For piece rate estimation and garment costing operation breakdown is made.

Sequencing of Stitching Operations

A garment is stitched by following the sequence of operations in a style and assembling all components. You cannot make the correct garment just by stitching garment components together randomly.

Sewing operators need to follow the garment construction sequence to make the desired product with correct measurement, correct fit and right shape. Industrial engineers prepare the garment operation bulletin and list down the operations in the best possible sequence that need to be followed when constructing the garment in a sewing line.

Considering you are a beginner in garment production, you will find helpful and valuable content in this piece of article. I have written this article in reply to the following question.

Question: *I am working in a garment factory that makes ladies fashion, semi-formal basic shirts. I am new to garment industry and am working as an industrial engineer. I am facing difficulty in making operation bulletin and preparing the operation sequence. Can you please provide me an operation bulletin for ladies' basic shirts, semi-formal shirts, and tunic?*

It is true that many young graduates join the garment industry without having prior training on preparing operation bulletin and they face such challenges at work in the initial days. You do not need to worry on this. You can learn how-to do things that you need to perform at your workplace, and you can master yourself on your job day by day. For achieving this, you must have right attitude for learning.

I have shared many operation bulletins of different apparel products on OCS blog. You can refer those sample operation bulletins for learning various kind of operations involved in different apparel products and possible sequence of operations within the style.

As you said, you are new to industrial engineering job and you do not have prior education on apparel production and garment manufacturing technology, you are at the learning stage. If I give you a couple of operation bulletins of the requested products, you will give less effort on your work in learning, you might copy it and use it in your factory by modifying some operations. That is not bad, but I want you learn things by doing it.

If you can make OB of a t-shirt and a woven shirt, you can also make the OB for the ladies' basic shirt/tunic/semi-casual shirt.

How to do that? You have emphasized on the sequencing of the operations in an operation bulletin.

You need to study the garment construction. If you have access to the sampling room in your factory, take help from the sampling team to learn the product construction for the upcoming styles. If the style is already running on the production floor, spend time on the floor and follow the sequence of operations performed by the operators to construct the complete garment. Operation sequencing is part of making an operation bulletin. Preparation of an operation bulletin involved two activities-

- Preparation of the operation breakdown - I mean making the list of all operations involved in making the product.
- Estimating SAM of garment operations.

Operation sequencing method

Follow the below steps to make an OB yourself.
- Take one sample garment
- Identify garment components. Like in a formal shirt garment components are - Collar, Cuff, front, back, and chest pocket.
- Prepare a list of operations in the sample garment component-wise. In case you unable to identify components, just list down operations.
- Group operations by garment components - if you find difficulty understanding any seams/operations, discuss it with the line supervisor or with sampling master. They can guide you in understanding the garment construction.
- Then think of the logical sequence of stitching the garment. Visualize how operators will make the garment. If there is more than one operation in a seam – what operation should come prior to another? If you need, open the garment seam, and look inside how the garment components are joined and possible sequence of operations inside the seam.
- Now place all operations in a sequence as per your understanding (do not worry if something goes wrong at the first attempt). Later, you can change the operation sequence by observing the process sequence on the shop floor.

You are done. See the following example of operation sequencing for making a full sleeve men's shirt.

S.N	MEN'S FULL SLEEVE SHIRT	
	Operation Break Down	SAM
	POCKET PREPARATION	
1	Mark pocket	0.288
2	Pocket mouth iron	0.353
3	Hem pocket (1)	0.457
4	Crease Pocket	0.528
5	Trim pocket	0.523
	COLLAR PREPARATION	
1	Mark Lining	0.235
2	Collar run-stitch	0.732
3	Collar turn & Collar Iron	0.331
5	Collar top-stitch	0.520
6	Collar Band Hem	0.533
7	Collar attach to band	0.800
8	Collar trimming, marking & notching	0.617
9	Collar band centre stitch	0.500
	CUFF PREPARATION	
1	Cuff hem	0.650
2	Runstitch cuff	0.800
3	Turn cuff	0.350
4	Iron cuffs	0.454
5	Topstitch Cuff	0.840
	SLEEVE PREPARATION	
1	Cut sleeve slit at plkt position	0.607
2	Notch Sleeves	0.510
3	Iron upper and lower sleeve placket	0.515
4	Attach Plackets	1.307
5	Close lower placket	0.865
6	Close upper plkt & make diamond	1.822
	BACK YOKE	
1	Make Pleats at back	0.482
2	Back yoke attach & Topstsitch	0.675
	FRONT PREPARATION	
1	Mark front for pkt position	0.188
2	Form Button hole plkt	1.117
3	Crease B/H placket (single fold)	0.256
4	Topstitch B/H placket	0.352
5	Sew button plaket	0.465
6	Attach pocket (1 pkt)	0.900
7	Sew label at placket	0.711
	ASSEMBLY	
1	Set front & backs & mark neck for collar	0.636
2	Shoulder attach	0.977
3	Shoulder top stitch	0.561
4	Sleeve Attach	1.391
5	Topstitch Armhole	0.709
6	Side Seam	0.841
7	Collar Attach	0.948
8	Collar Close & insert label	0.780
9	Cuff attach & close	1.382
10	Bottom Hem	0.926
11	Button Hole - Plkt & Collar	0.500
12	Button Attach	0.500
13	End of line checker	1.250
	TOTAL SAM	**31.684**

Why correct operation sequencing important?

An operation bulletin with correct operation sequence helps in the following ways.
- Understanding garment construction
- Understanding process flow of making a garment
- Production line setup with correct operation flow
- Easy to display line balancing
- Managing bottleneck in the production line

For the other parts of the OB making, you need to learn some calculation methods other than estimating operation SAM. Those are - estimated manpower, estimated machine requirement, hourly target, actual manpower to be allocated depending on the daily production requirement and line balancing. I have elaborated these calculations in the next article *Making an operation bulletin of a garment.*

Making an Operation Bulletin of a Garment

An operation bulletin (OB) is one of the primary IE tools. An OB helps in setting up a production line with the correct number of machines and manpower. To make it easy for learners, I have explained OB making process step by step. Prior to using these steps for making an OB, you need to know how to establish operation SAM, calculation method of machine requirement, and how to make operation breakdown of a garment. You will learn these here.

			Operation Bulletin (OB) Template								
Style		#35427	Style Description		Shirt	Buyer			SST		
Target Eff%		60%	Machine SAM		27.808	Order Qty.			50000		
Shift Time		480	Manual SAM		3.23	Date			22-Apr-20		
Daily Target		800	Total SAM		31.034	Engg. Name					
Seq. #	Operation Code	Operation Description	Machine Desc.	SAM @ 100%	Prodn capacity / Hr. @ 100%	SAM @ 60%	Prodn per m/c/ Hr. @ 60%	Folders & Attach ments	No. of Calculated M/C @ 60%	No. of machines allocated	Total Prod./hr @ 60%	Remarks
		Collar Making										
1	110	Mark Lining	Manual	0.235	255	0.39	153		0.65	1	153	
2	120	Collar run-stitch	SNLS	0.732	82	1.22	49		2.03	2	98	
3	130	Collar turn & Collar Iron	Manual	0.331	181	0.55	109		0.92	1	109	
4	140	Collar top-stitch	SNLS	0.520	115	0.87	69		1.44	2	138	
5	150	Collar Band Hem	SNLS	0.533	113	0.89	68		1.48	2	135	
6	160	Collar attach to band	SNLS	0.800	75	1.33	45		2.22	2	90	
7	170	Collar trimming, marking & notching	Manual	0.617	97	1.03	58		1.71	3	175	
	180	Collar band centre stitch	SNLS	0.500	120	0.83	72		1.39	1	72	

Figure-1: Operation Bulletin Template

Step-1: Prepare an OB format.

You need a computer with a spreadsheet (Excel sheet) to make this format and to do calculations. In the Figure-1, I have shown one sample operation bulletin template. I will explain steps using this example. You can copy this format for your use or make one yourself. On the header, add details like style#, Buyer, Order Qty. etc.

Add a formula to the cells for auto calculation when you enter operation SAM (in the 5th column). The formula for each column has been given in the following steps. A basic operation bulletin contains the following information.
- Daily working hours (shift in minutes)
- Target output per day or Target per hour
- Total SAM (Sewing SAM + Non-sewing SAM)
- OB prepared by (Name of the Engineer)

- Operation code (Optional)
- Operation Name/description
- Machine Name/description
- SAM @ 100% Efficiency
- SAM @ target efficiency%
- Calculated production per hour
- Folder and attachment name if used
- Calculated number of machines
- Actual number of machines
- Estimated production per unit hours
- Remarks
- Machine summary list (at the bottom of the OB sheet)

Step -2: Collect the correct sample

Once you have OB format ready collect one style (reference sample) from sampling or merchandising team. When you collect a sample from sampling room or from a merchandiser, make sure that you take an approved sample.

Step -3: Make operation breakdown for the sample

Look into garment construction in details. To have better understanding first break garment into parts (like a collar, front, back etc.). Then list down operations in your notebook according to the garment component. Once you list down all operations, cross-check with garment sample again and visualize whether all operations are added as per your sample garment. Now enter operations in the OB format (computer spreadsheet) following the operation sequence to be performed in the line.

Step -4: Identify machine for each operation

Based on seam type used in a style you may need to use different types of sewing machines for different operations and manual workstations. Select one machine and enter machine name in the sheet against each operation. Also, enter name and description of attachments or folder or guides if needed for an operation.

Step -5: Enter SAM for each operation

This step is the most critical in term of how you get the SAM of each operation. You can establish SAM by conducting time study or use SAM from the existing database. Mention the source of SAM (time study or MTM2 database). Convert SAM based on the target line efficiency%. It is optional. In the example (OB template), SAM is shown at 100% efficiency and converted SAM into Target efficiency (i.e., 60%). Refer to the formula.

> SAM at Target Efficiency = (SAM @ 100% efficiency / Target efficiency%)

Step -6: Calculate production capacity per hour

Calculate production capacity per hour at target efficiency%. In our example, I have shown production capacity at 100% and Target efficiency (60%).

> Production/Hour @ Target efficiency = 60/Operation SAM @ Target efficiency

Step -7: Calculate number of machines

To make garments as per daily production target, you need to arrange enough sewing machines. You will get fraction number of machines (with decimal) in this calculation. In the next column add the number of machines manually round figure for machine numbers. Combine two operations in a single machine where possible, i.e., required machine number is less or equal to half.

> Calculated Machine number = (SAM @ Target efficiency x Hourly production target/60)

You can also use pitch time for the calculation of calculated machine number. Formula:

Calculated Machine number = (Operation SAM / Pitch)

Step -8: Calculate estimated production per hour

It is calculated according to the machine assigned to each operation. It may be different for few operations than calculated production per hour (shown in Step -6) as you set machine numbers to round up value.

> Estimated Production/Hour = (No. of machine assigned x Hourly target qty.)/Calculated no. of machines

Step -9: Calculate machine summary

At the bottom of the OB add one table for machine summary. This table will help you to quickly find number of machines required for each machine category. In this table, add formula to calculate machine summary automatically when you enter the number of machines in the OB.

Machine Summary			
Machines	SAM	No. of Machines (calculated)	Actual M/c
SNLS	8.07	18.69	19
4T O/L	0.86	2.00	2
DNLS	0.46	1.06	1
Fusing M/c	0.13	0.30	0
BS	0.30	0.70	1
BH	0.29	0.67	1
5T O/L	1.90	4.39	4
Flatlock	0.64	1.47	1
MANUAL	0.40	0.93	1
	13.05	30.20	30

Conclusion:

The operation bulletin template can be designed many ways depending on the information you need to show in the OB. You just need to ensure that all necessary information must be included in the OB template.

A basic format of an OB template is shown in the next page (Figure-2: Sample operation bulletin for a full sleeve formal shirt).

Operation Bulletin (Full Sleeve Formal Shirt)

Parameters		
Style	Mens Shirt	
Output (pieces per day)	50	
Target Efficiency (Start up)	50%	
No. of Workplaces (Calculated)	8	
Shift time (minutes)	480	
Absenteeism	10%	
Operators - Sewing		
Total SAM	33.44	
Pieces/WS/ day	6	

S.No.	Operation	SAM (Standard Time)	Machine Type	No. of operators (calculated)	No. of m/cs Calculated.	Comments
	Collar Making					
1	Mark Lining	0.24	Manual	0.05	0.05	
2	Collar run-stitch	0.73	SNLS	0.17	0.15	
3	Collar turn & iron	0.33	Manual	0.08	0.07	
4	Collar top-stitch	0.52	SNLS	0.12	0.11	
5	Collar Band Hem	0.53	SNLS	0.12	0.11	
6	Collar attach to band	0.80	SNLS	0.19	0.17	
7	Collar trimming, marking & notching	0.62	Manual	0.14	0.13	
8	Collar band centre stitch	0.50	SNLS	0.12	0.10	
	Cuff Section			0.00	0.00	
9	Cuff hem	0.65	SNLS	0.15	0.14	
10	Runstitch cuff	0.80	SNLS	0.19	0.17	
11	Turn and Iron Cuff	1.00	Manual	0.23	0.21	
12	Topstitch Cuff	0.84	SNLS	0.19	0.18	
	Pocket Section			0.00	0.00	

Figure-2: Sample operation bulletin

The Secret of Making Operation Bulletin for Beginners

Making an Operation Bulletin (OB) is not a big deal. If you understand sewing machines and garment operations and can work on the Excel (spreadsheet), you can make an operation bulletin.

If you are starting your career in a garment factory as an industrial engineering, you must know the secrets of making operation bulletin and estimating SAM for the different garment operations. Remember, you are not going to make the perfect OB in the initial attempts. You need to practice a lot to become master in making the operation bulletin.

Also remember, the most difficult part of making OB is estimating operation SAM (Standard minute). If your factory has SAM calculation software (like TimeSSD, GSDCost, SewEasy or ProSMV), learn how to use the software. If you do not have any software, do not worry. As in a recent survey, I found that less than 26% of garment factories use PMTS based software for calculating garment SAM. Rest 74% of companies do not have such system but they make operation bulletin with estimated SAM and meet the purpose of making OB. They do machine and manpower calculation, initial line balancing based on the estimated SAM and run the factory well.

There is a solution for this problem too - use **Time Study** method for establishing garment SAM. The secret formula is explained in the following pointers. Practice using following guidelines and become an expert in making OB.

1. Collect a few operation bulletin sheets from the IE department. If you do not have access to the IE department in your factory you can refer a couple of operation bulletins published in my blog *www.onlineclothingstudy.com/search/label/Operation%20bulletin*.
2. Read the existing OB thoroughly and try to relate operation and operation SAM at 100% rating. Go through the number OBs of same products and cover multiple styles for the same product.
3. Create one OB template in the spreadsheet following sample OB. Add formula in machine and manpower column. If you do not have an OB format, download <u>sample OB sheet from OCS blog</u>.
4. Take one garment (sample) and make operation breakdown.
5. Match operations of the sample garment with example OB and use that SAM in your OB. Fill SAM for operations as many as

possible. In case you do not find data in the sample OBs, keep it blank for time being.
6. Wait for loading the style to the line. Conduct cycle timing at least 15-20 consecutive operation cycles for each operation. To get accurate result first study sewing operator - if the operator is doing operation in a rhythm (neither at slow pace nor at fast pace). After doing time study for few weeks, you will be able to find operators working with normal pace.
7. From the cycle time, calculate standard minute for sewing operations. Fill the blank cells in your OB with SAMs calculated through time study.
8. Compare pre-filled SAM of all operations with the average cycle time. Check whether they match or there is a big variation. If you find a big variation, study another sewing operator. If you still find there is a big gap between assumed SAM and Time Study based SAM, replace the SAM in the OB with time study value.
9. Other sections(columns) of the OB sheet can be calculated automatically by setting formula in the spreadsheet cells.
10. When you complete making a dozen of OBs by following the above method, you will be comfortable with making an OB for similar operations. For new products and new operations, follow above method again. Afterword you will be making garment OB and estimate operation SAM based on your experience.

Why use experience-based SAM?

In my first interaction with an industrial engineer in a Delhi-based export house in 2005, I asked her where she got those standard times (SAM) written on the operation bulletin. She replied that they have taken standard time for each operation based on their experiences. They were not using any tool/software for SAM calculation – SAM values are an assumption based - as they are experienced, they know what time it should be. Though there might be a question of the accuracy of SAM values.

Things have not changed much in last 15 years. As stated above many factories determine garment SAM based on their experience. Why not you follow the same method?

The No.1 secret for calculating standard time for garment operation is calculating garment SAM. You need to gain experience by doing the job - by preparing OB and observing operators in the sewing floor and conducting time study as much as possible.

Line Layout

You know, the garment stitching is the core process in the apparel production and in the fashion supply chain. Many garment manufacturers use the assembly line for making garments for mass production.

In an assembly line, number of sewing machines (including different types of sewing machines and non-sewing equipment) are placed in a line according to the process sequence requirement. Garment bundles are loaded at one end of the line and moved from one workstation to another, and finally stitched garments come out from the line.

Definition of Line Layout

When we talk about preparing a **line layout**, it means designing the presentation of workstations in an assembly line and showing the flow of work from start of the line to end of the line. It can be a simple line diagram. See the righthand side diagram. The rectangle boxes are indicating sewing workstation. The number marked inside the boxes are indicating operations sequence in the operation bulletin and arrows are showing the flow of work. The line layout is normally prepared by IEs after making the operation bulletin.

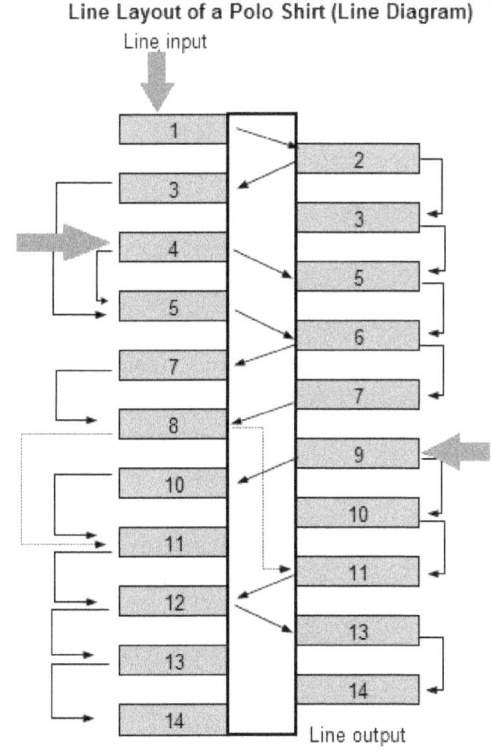

Line Layout of a Polo Shirt (Line Diagram)

Line Layout of a Product (Particular style)

There are different types of sewing line layouts. Whatever the line layout one factory uses, industrial engineers need to prepare line layout for the upcoming orders.

For an example, a production line involved different types of sewing machines, non-sewing equipment, and workstations for manual processes. Each workstation is designed for an operation.

When you prepare the line layout, you will see following cases and you will define a workflow on the line layout.

- You may need to assign more than one workstation for the same operation,
- You may need to use same machine (workstation) for more than one operations
- You may need to load bundles or garment component in multiple workstations

Why prepare a line layout?

Followings are some of the primary benefits and applications of line layout.

- The line layout helps line supervisors to set the line for a new order in the least time.
- From the layout, you will get the exact requirement of machines and equipment. Production team can be better prepared.
- Style changeover time can be reduced by preparing a detailed line layout and doing line set-up based on the line layout
- For new plant set-up, designing the line layout is an essential part. Looking at line layout, overhead wiring, busbar installation, compressed-air connection, installation of ironing tables, QC tables and fixture are installed.

In case a factory produces same product for all the time, still they need to change the line layout depending on the product design and operation sequence. In such cases, factory may make minimum changes of sewing machines and workstation layout.

What is the Balancing Loss?

Prior to loading cuttings to a production line (for a new style), industrial engineers prepare an operation bulletin (OB) for the style. In the operation bulletin, IE plans calculated manpower and machine requirement based on SAM of all operations (pitch time) and production target per hour (or per day). This is known as calculated (or theoretical) machine requirement. In the calculated machine numbers, a fraction of machine number may come but practically fraction of machine cannot be set in the sewing line.

In the operation bulletin, an IE also plans the actual machine requirement for each operation. Where possible, operations are clubbed (where similar machine type is required) for balancing available time per machine and SAM of operations. But for all operations balancing is not possible. That is why in the OB, IE gets variation in between two figures - calculated machines and actual machine numbers. Total variation of the machine numbers for a style is known as **balancing loss**.

Let us take an operation label attaching using computerized single needle lock stitch machine. Compared to other operations, SAM of this operations is low. As a result, target production is achieved in less than given hours. So, for this machine there is no work for full day. This machine cannot be used for other operations. So, the line will lose operator's productive time.

For an operation (e.g., label attaching), balancing loss can be reduced by increasing production target. But to meet the increased target in one operation, one may loss balancing in other operations.

See in the following sample operation bulletin of a Polo Shirt. Calculated machine number and allocated machine numbers are respectively 26.42 and 29. Variation of machine numbers is 2.58 machines.

Therefore, balancing loss for this sample style is 9.78%

How to Calculate Balancing Loss?

The calculation procedure of balancing loss% for a style is explained here.

Calculate machine number based on operation SAM and production target per day. For detailed procedure of calculating machine requirement in the OB, read the article *'How to calculate machine requirement'* shared in this book.

Industrial Engineer's Digest

Table - Operation Bulletin of a Polo Shirt

	Style	XYZ				
	Output (pieces per day)	800	Shift	480	SAM	12.68
	Target Efficiency	80%	Absenteeism	10%		
	No. of Workplaces	29	Machine	29		

Sl. No.	Operation	SAM	Machine type	No. of operators (calculated)	No. of m/cs Calculated.	M/c allocated
	Preparatory					
1	Make tape for side slits	0.20	Flat lock	0.46	0.42	1
2	Placket Folding by iron	0.60	IRON	1.39	1.25	1
	Front					
1	Attach top and bottom placket to front.	0.53	SNLS	1.23	1.10	1
2	Counter stitch placket	0.24	SNLS	0.56	0.50	1
	Assembly					
1	Shoulder Seam	0.36	3TOL	0.83	0.75	1
2	Flat Lock on Sh/Seam	0.47	Flat Lock	1.09	0.98	1
3	Collar Attach	0.37	3TOL	0.86	0.77	1
4	Neck Binding	0.57	SNCS	1.32	1.19	1
5	Neck Finish	0.50	SNLS	1.16	1.04	1
6	Placket finish	1.10	SNLS	2.55	2.29	2
7	Placket edge OL	0.11	3TOL	0.25	0.23	
8	Sleeve Attach	0.80	3TOL	1.85	1.67	2
9	Flat Lock on Arm hole	0.90	Flat Lock	2.08	1.88	2
10	Side Seam + Under arm	0.89	3TOL	2.06	1.85	2
11	Sleeve Hem	0.76	Flat Lock	1.76	1.58	1
12	Bottom Hem	0.64	Flat Lock	1.48	1.33	2
13	Attach Tape on slits	0.69	SNLS	1.60	1.44	2
14	Finish slits	1.10	SNLS	2.55	2.29	2
15	Tack on Sleeve hem	0.30	Bar tack	0.69	0.63	1
16	Tack on Side slits	0.30	Bar tack	0.69	0.63	1
17	Button Attach (3 Nos)	0.45	Button Atta.	1.04	0.94	1
18	Buttonhole (3 Nos)	0.45	Buttonhole	1.04	0.94	1
19	Label attach	0.35	SNLS CP	0.81	0.73	1
	Total SAM	**12.68**		29.35	26.42	29.00
			Machine Variation			2.58
			Balancing Loss			9.78%

(Note: In the above example absenteeism percentage is considered to find number of calculated machines. You can exclude absenteeism in your calculation.)

Allocate machine to each operation. Only enter nearby round number for allocated machine. (The computed machine is rounded to the nearest integer. Any value less than n.50 is rounded down to the integer "n". Any value equal to or greater than n.50 is rounded up to the integer "n+1". For example, calculated machine number 1.25 results in 1 and 1.70 results in 2.

For the critical operation, you might add machines to next round number. Where combined SAM of two operations is below or nearby pitch time, you can consider one machine for those two operations. This way, enter allocated machine numbers

Find the difference between the number of allocated machine and number of calculated machine.

Calculate balancing loss% by using the following formula

> Balancing Loss% = (Number of allocated machine – Number of calculated machine)/Number of calculated machines

An example of balancing loss calculation is shown in the above operation bulletin. Where,

Calculated machine number: 26.42 and Allocated machine numbers: 29.

Variation of machine numbers is = (29-26.42) = 2.58 machines.
Balancing loss%= (2.58/26.42) x 100 =9.77%

(PS: You may see difference in balancing loss percentage in the image (9.78%) that is due to round off the decimal point)

Sewing Process Flow Chart of a Garment

It is always easier to understand a picture and flowchart than written description. How to make a t-shirt can be explained many times but students learn it completely when they see the t-shirt making process in the production floor. A process flow chart gives them the logic how operations follow sequence one after another.

Why garment sewing process flowchart is prepared?

The process flow of a garment helps operators to construct the garment (assembling garment components) in a correct sequence. When stitching a garment, they do not need to waste time in thinking about in what sequence garment component should be assembled. In mass production, the process flowchart helps industrial engineers and line supervisors in setting up the line for machine layout.

A sewing process flow chart is depicted in the following Figure-1 to guide you how a garment (T-shirt) is being made in the bulk production system. It is assumed that T-shirt has neck tape (fabric bias cut tape). The process flow is also showing the sequence of operations that are generally being followed. Some factories may have a slightly varied sequence.

A crew neck T-shirt has six components – Front, Back, Neck rib or Collar, Neck tape and two sleeves. To make the sewing process flow chart, first make operation breakdown following operation sequence and list down all operations in a t-shirt.

Seq. no.	Operations
1	Shoulder join
2	Make collar rib
3	Attach collar rib to neck
4	Set neck tape
5	Down stitch neck tape
6	Tops stitch neckline
7	Attach sleeve to armhole
8	Side seam and underarm overlock
9	Hem Sleeve and Hem bottom
10	Attach care label and main label

In the flowchart, on the top, four parts (front, back, sleeve and collar) of a t-shirt is shown. The arrows show the flow of operations and inside the circles, the operation sequence number and the name of the operations are written.

Figure: Sewing process flowchart of a T-shirt

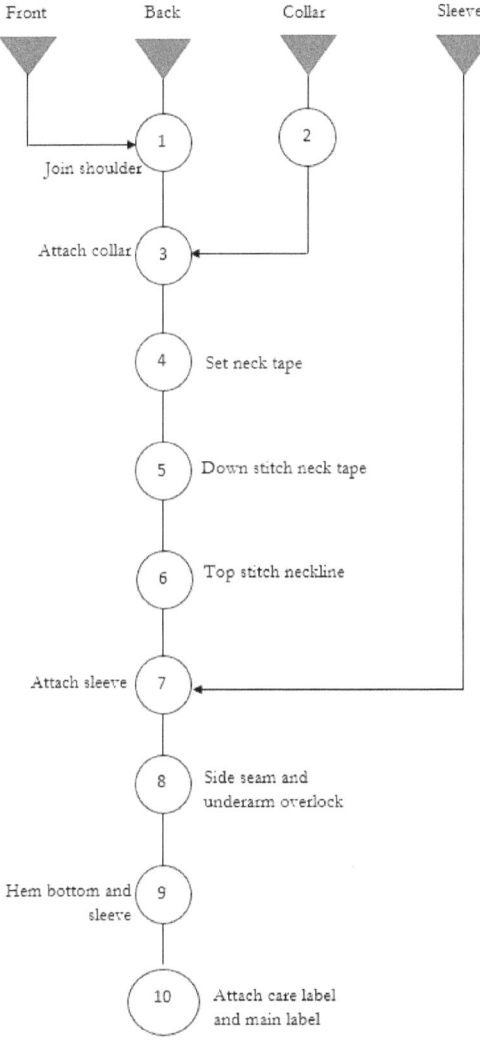

By following the above method, you can prepare process flowchart of any other garment styles. Before you make the process-flow chart, you need to identify garment components, prepare the operation breakdown, and identify the operation sequence of making the garment.

Sewing Machines Used in Making Polo Shirts

Do you know what types of sewing machines do you need for making a polo shirt style?

If you already know the types of sewing machines used for making a polo shirt and machine requirement calculation method, you can skip this article. This is for newcomers. A ready reference of machine types is shown here for making polo shirts.

Let's say, you are going to make polo shirts in your factory. You need to identify machine types from the construction of a polo shirt. Normally, five different types of sewing machines are used in making basic polo shirts.

1. Single needle lock stitch machine,
2. Over edge (Overlock),
3. Button holing,
4. Button attaching and
5. Flat lock (Flatbed).

Type of sewing machines used for performing different operations in a polo shirt are shown in the following table. First prepare the operation breakdown following operation sequence and then following the machine calculation method find operation wise machine requirement for given daily production target.

Number of sewing machines needed for making daily target production quantity is calculated based on operation SAM. Let's say, daily production requirement is 500 pieces in 8 hours shift. When calculating machine requirement, some machines can be shared for multiple operations.

Table: List of operations and types of sewing machines

Seq. No.	Operations	M/c Type	No. Of Machines
1	Placket Rolling	1N Lock Stitch	1
2	Placket Join	1N Lock Stitch	1
3	Nose Tuck	1N Lock Stitch	1
4	Shoulder Join	Over lock	1
5	Collar Join	1N Lock Stitch	2
6	Collar Piping	Over lock	1
7	Upper Placket Top	1N Lock Stitch	1
8	Lower Placket Top	1N Lock Stitch	1

9	Back Neck Top	1N Lock Stitch	2
10	Placket top	1N Lock Stitch	1
11	Placket Box	1N Lock Stitch	2
14	Sleeve Cuff Join	Over lock	1
15	Cuff Top Stitch	Flat lock	1
16	Sleeve Join	Over lock	1.5
17	Arm Hole Top	Flat lock	1
18	Side Seam	Over lock	1.5
19	Body Hem	Flat lock	1
20	Sleeve tuck	1N Lock Stitch	1
21	Buttonhole	Button Holing	1
22	Button Stitch	Button Stitching	1
	Total		24

1N lock stitch stands for single needle lock stitch machine

Remember, the machine types and number of machines mentioned here are based on some variables like product design, product SAM, daily target, operators' absenteeism rate, line efficiency percent.

In the next article, I have shared a list of sewing machines required for making a t-shirts in mass production environment.

Sewing Machines Used in Making T-Shirts

To make quality t-shirts with good finishing inside and outside of the garment, normally three sewing machine used - overlock machine (for edge stitch), flatlock (for cover stitch) and single needle lock stitch machine. But for decorative stitch other sewing machines can be used.

For your ready reference, type of machines used for sewing different operations and machine requirement for making 800 t-shirts daily is shown in the following table.

When calculating machine requirement, some machines can be shared in multiple operations.

Table: List of sewing operations and sewing machine types

Seq. No.	Operations	M/c Type	No. Of Machines
1	Shoulder join	4 Thread Over Lock	1
2	Neck rib Tuck	1N Lock Stitch	1
3	Neck Join	4 Thread Over Lock	2
4	Label Make	1N Lock Stitch	1
5	Back Neck Binding	Flat lock/1NLS	1
6	Front Neck Top	Flat lock	1
7	Back Neck Top	1N Lock Stitch	2
8	Sleeve Hem	Flat lock	1
9	Sleeve Join	4 Thread Over Lock	2
10	Side Seam	Flat lock	2
11	Sleeve tuck	1N Lock Stitch	1
12	Body Hem	Flat lock	1
	Total		16

Machine summary:

To make 800 pieces of t-shirts daily, you need 5 overlocks, 5 single needle lockstitch and 6 flatlock seeing machines. If you share the lock stitch machines for multiple operations, 3 lockstitch machines will be enough for making 800 t-shirts daily.

Remember, when calculating number of machines for making a t-shirt design some variables are considered -like product SAM, daily production target, operators' absenteeism rate, line efficiency percent.

Line Balancing

"Line balancing means leveling the workload across all operations in a production line, removing bottlenecks and utilizing excess capacity of operators", defined by Six Sigma Material.

When you consider mass production, garments are produced in the production lines or using a set of machines, instead of single sewing machine. A line may be an assembly line, modular line or section, or a line set-up with online finishing and packing. A line includes multiple workstations with varied work contents.

Production per hour from an operation varies depending on the work content (standard minutes of a task/operation), allocation of total manpower to an operation, operator skill level and machine capacity. Operation with the lowest production per hour is called a bottleneck operation for that line.

A bottleneck operation in a line restricts the output of the line. That is why it is very important to increase production of the bottleneck processes and bottleneck operations.

Why is line balancing done?

Line supervisors and work-study officers find ways to increase production from the bottleneck operation and implement those means one by one to level work across operations in a production line. In layman language, this is called line balancing.

Secondly, Line balancing is essential as because, if an excess capacity of sewing operators does not utilize, production cost will be high and results in waiting and absorption of fixed cost.

Though above definition is widely accepted, I have seen a few factories where industrial engineers name line balancing to something else. At the time of machine/manpower planning is done based on work content of each operation, they prepare a sheet in which operation wise manpower is calculated. Operation wise manpower planning and machine planning to produce the target quantity is called as line balancing. This is also called as initial line balancing.

Most of the cases calculated manpower gives a fraction of value, but practically you cannot allocate the fraction of manpower to an operation. Therefore, manpower planner decides to which operations one machinist, to which operations two machinists or where only single machinist will be allocated for two or three operations. Planner makes this decision based on calculated data.

To view how your production line balanced, you can draw a line chart with production data and display hourly production from all the operations (or only from critical operations). In the line chart, you can see the bottleneck operations. In the following chart, production per hour of two lines are shown. Following line chart indicates that Line B is better balanced than Line A. Further, you can compare operation wise hourly production with the hourly production target of the line.

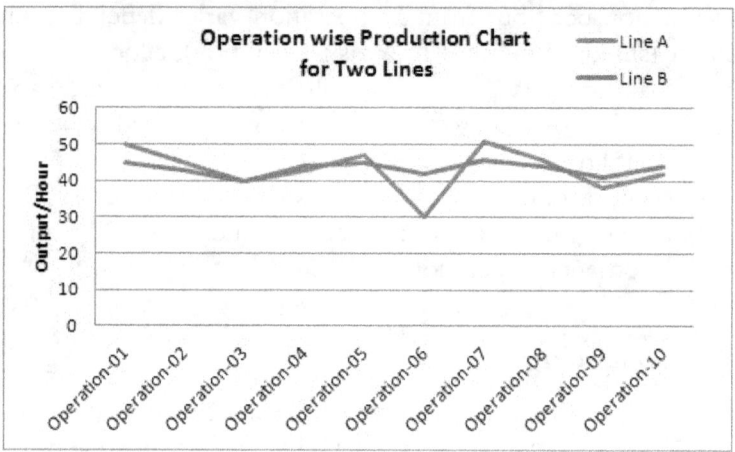

Figure 1: Line Balancing Graph

The above example shows the importance of line balancing in improving line's production. Later, in another article, I will show you how to balance a traditional sewing line.

Different Types of Garment Production Systems

In laymen's language, a 'garment production system' is the way how the fabric is being converted into a garment in a manufacturing system.

Production systems are named according to the various factors, like- number of machines is used to make a garment, machines layout, total number of operators involved to sew a complete garment and number of pieces moving in a line during making a garment. As the fashion industry evolved and demand for readymade garments are increased, the need of mass production systems become the essential way to meet the global market demand. Simply because the tailoring shops are not capable to produce high volume orders and supply across the world.

Different Types of Garment Production Systems

The most common production systems found in the garment industry include the followings
- Make through system
- Progressive bundle system (PBS)
- Section production system
- Modular production system
- One-piece flow system
- Overhead production system

Each of these production systems are explained briefly in the following sections.

Make Through System

When a tailor alone makes a complete garment, it is called as make though systems. The tailor even makes a pattern (uses ready-made patterns), cuts fabric and does the finishing of the garment. For example, tailors in the tailor shops do all jobs from cut to pack. In this system, a tailor is not depended on the other tailors.

In a garment factory, the sampling room follows the make through production system where one single operator makes the whole garment sample. I have seen make through system is followed in the high fashion garment making factories and, in the factories, where they used to make garment for make-to-order model. In a make through production system, all components of a garment and necessary trims are packed in a polybag and given to the operator.

Progressive Bundle System

In progressive bundle system (PBS), each operator does different operations of a garment. All sewing machines needed to make the garment are installed in a line. Garment cut parts are fed to a line in a bundle form. When an operator receives a bundle of cut components, she opens the bundle and does her operation (job) for all pieces of the bundle. After completing her job, she moves the bundle to the next operator who is doing the next operation.

Several operators and helpers are involved in sewing a single garment. Major benefits of this system are – as operators work on single operation throughout the day, they gain skill on the given task. As a result, the line works at a better performance level.

Secondly, product consistency can be maintained garment to garment. Most of the export-oriented garment manufacturers adopted progressive bundle system as the main production system.

Section Production System

This system is like the progressive bundle system. But the difference is that, instead of one line, work is divided into sections. Machines of similar operations are clubbed together instead of spreading over in all lines. For example, when a man's formal shirt is being made in a section layout – collars, cuffs and sleeves are made in the preparatory sections and then ready parts are sent to the assembly section. This system is popular to improve line balancing and utilization of human resources.

Modular Production System

In 'Modular production system' sewing operators work as a team. Neither they sew complete garment, nor do they sew only single operation. Multi-skilled operators form a group and each of the team members do multiple operations.

In a modular system, operators help each other to finish the garment quickly and the team is fully responsible for quality and production. In modular, always team performance is measured instead of individual operator performance. This system is very successful where quick response is needed.

One Piece Flow System

In one-pieces flow system, single-piece bundles are moved from one operator to the next operator in an assembly line. Instead of making a bundle of multiple pieces, a bundle is made of all components of a single piece. Sewing machines in one-piece-flow system can be laid in a straight line or modular line. The main difference is that the operator will receive one piece from the back and move one piece to his next operator after completing his work. Benefits of one-piece-flow system are less throughput time, less WIP in the line.

Overhead Production System (UPS -Unit Production system)

In the overhead production system, garment components are clamped in a hanger and the hanger moves on an overhead rail. In the hanger, components of a single piece are clamped. So, this is also one kind of single-piece-flow system. The benefits of overhead material handling system include – reduced material handling time during stitching garments, no need of bundle tying and bundle untying.

Whatever Production system is used as mentioned above, when operators are paid according to their works (how many pieces produced), is named as piece rate system.

Conventional line Vs Modular line

In the garment manufacturing industry, various production system is found, and each production system has some benefits over others depending on the order size, product type, technology level and skill level of the operators. When it comes lean manufacturing system in apparel manufacturing, factories start implementing modular system whatever product they might produce. It is true that in a Lean environment, the modular line is the most effective one than a conventional line if the modular line is run scientific way. Here conventional line means a progressive bundle system which is also known as the assembly line.

To support the above statement, I will take example of Madura Clothing's shirt line. In a conference (OGTC conference, 2011) Mr. Lal, Head of Manufacturing, Madura Clothing, presented a case of the Shirt line. They have both a conventional line and Modular line on the same floor.

According to his presentation, in a modular line, they produce 14.3% more shirts with 23% less sewing operators comparing to the conventional line.

Secondly, the labor productivity (operator only) of the modular line is 48.6% higher than the conventional line.

In the following table, comparison has been shown with other parameters.

Table: Conventional system Vs Modular production system

	Parameters	Conventional Line	Modular Line	Total of 2 Modular line	Improvements
1	Production/day	700	400	800	14.3% Higher
2	No. of Operators	52	20	40	23.1% Less
3	Takt time	35	70	70	Higher
4	Throughput time (Theoretical)	1820	1400	1400	Less
5	Effective Throughput time	1870	700	700	Less
6	Labor Productivity	13.5		20	48.6% Higher

The above data shows the modular line yields better productivity in the shirt manufacturing.

I am sure, you will be also using one or both types of the production systems in your factory. You can use these matrices to compare performance of two different production systems. In the next article, I have shared the comparative analysis of progressive bundle system and overhead production system.

Progressive Bundle System Vs Unit Production System (UPS)

In the readymade garment manufacturing plants, various types of sewing systems are installed by different garment factories. A plant owner chooses one of these systems depending on the production volume, product categories, and cost effectiveness of high-tech machines.

Among those "Progressive Bundle System" (PBS) is mostly installed sewing system till date. In this production system, bundles of cut pieces (bundle of 5, 10, 20 or 30 pieces) are moved manually to feed the line. Then inside the line, an operator himself drag the bundle from side table and transfer the bundle to the next operator after completion of the work.

With the advancement of the technology, mechanical material transportation systems are brought in the sewing plant. An overhead material transport system, known as UPS (Unit Production System), transports cut pieces clamped and hanged in hangers (one hanger for one piece) by automated mechanical transport system. It reduces manual transportation, and it has many other benefits compared to the progressive bundle system.

This article is not intended to recommend you replace the well-placed progressive bundle system. When to install a new technology depend on various factors.

A comparison between these two production systems has been drawn in the following table based on production KPIs (Key Performance Indicators) to show you how an UPS system (overhead hanging and sensor-controlled system) is most effective over PBS.

Parameters	Progressive Bundle System (PBS)	Unit Production System (UPS)
Transportation	-Manual transportation, many times helpers are hired for this bundle transportation job. -Operators stop their work to fetch bundles. -Less effective in terms of production management. Resulted long response time.	-In this system an automated mechanical system carries pieces to each workstations. -Easier pick up and dispose at each workstation. Resulted quick response time
Throughput time	- Compare to UPS, throughput time longer in PBS. How much long will depend on the bundle size and no. of bundles kept in between two operators.	-Throughput time in UPS is less compare to PBS. But it is not the minimum time as in this system there is WIP in between two operators.
Direct Labor content	- Direct labor content is high because usually operator does tie and untying of bundles, positioning components, pulling the bundle ticket, and handling of work pieces.	-Direct labor content is less than PBS because an operator only sews the garment part rather than other tasks. In this system garment parts are held by the overhead hanger, so less handling of garment components.
WIP level	- In PBS generally operators are asked to sew as much pieces as they can without considering back and front operators. This resulted piling up of work in the operations with higher work content.	- Less WIP in between operators. As workstation has limit of holding no. of hangers. After completion of operation hangers are transported to the next operation automatically.
Cutting work requirement	- As a result of High Work in Process (WIP) is required by sewing section, cutting sections are required to perform 60-70% more than actual production can handle.	- Lower WIP results in less cutting works. A balanced flow of material established in between cutting and sewing line.

75

Industrial Engineer's Digest

Inventory Level	- Due high WIP and higher cutting, fabrics and trims need to stock in advance	- Less inventory for fabric and trims.
Excess labor requirement	- Usually in PBS needs more overtime works, repair work due to some unfinished operations.	- Plant with UPS system needs less overtime as planning is easy in this manufacturing system.

P.S.: The comparative study data and information collected from various articles and online resources.

Benefits of Single-Piece Flow Production System

Single-Piece Flow system is one of the many garment production systems. In Single-Piece Flow system only one garment is processed at a time through various operations in a sewing line. This production system is also known as a one-piece flow system. A single-Piece flow system is considered as a part of the lean manufacturing system. So, lean philosophy is applied when one uses a single-piece flow system.

Like other production systems, this production system has several merits and demerits. Benefits explained here are irrespective of lean manufacturing.

Reduces throughput time - Time required coming out pieces from the production line after feeding is quicker than traditional progressive bundle system. Reduced throughput time of an order has many other subsequent benefits.

Quicker response – This system reduces overall lead time for an order processing. Factories can respond to market demands quickly.

Faster money circulation - In this system, not much inventory is stored, and production lead time is less. The factory does not need to block money in a lot of material purchasing, transportation and warehouse.

Easier to change product - Setting up a line for new order or a new style is easier in the one-piece flow system. Style change over time is also less compared to progressive bundle system as there is not much WIP in the production line.

Improves garment stitch quality - Due to less throughput time, faults are detected quickly and feedback on stitching quality is given faster for preventive action. Also, in lean manufacturing one-piece flow system, quality is in-built. This means each operator is responsible for checking the quality of the work. No defects passed to the following operations. This way defect generation at machines can be reduced. Secondly, when there are fewer defects in garments, quality checker checks a garment in less time compared to existing production system.

Reduces rework level – Fewer defects mean less rework.

Better quality control – In this system, as any fault is detected within a minute of its occurrence garment quality can be managed in a better way.

Improves labor productivity – Productivity in a single-piece flow system is higher than a batch production. Reason – Non-value-added activities are eliminated in the one-piece flow system. Thus, an operator can produce more standard hours in the given time.

Better line balancing –Equal work distribution to operators can be done by clubbing low work content jobs. As a result, minimum bottlenecks exist in the line.

It is considered that the cost of manufacturing for smaller quantity order is less in a single-piece bundle system compared to the progressive bundle system.

Different Types of Industrial Sewing Machines and Their Use

Industrial sewing machines are normally used in mass garment production. An industrial sewing machine is power-driven and runs at a very high speed. There are different types of industrial sewing machines. There are some special sewing machines developed for making specific seam and stitch classes.

To have in-depth knowledge of sewing machines, learn about different types of sewing machines. Application of these machines is also mentioned with an example. This would help you visualize the machine's application in making the common products.

1. Single Needle Lock Stitch Machine

This machine makes lock stitches (stitch class 301). Lock stitches are formed with one needle thread and one bobbin thread. This is a widely used sewing machine and used for sewing stitch class 301. Basic to computer-controlled version are available in this machine category. In the single needle lock stitch machine, there are machines with underbed trimmer (UBT) and without UBT.

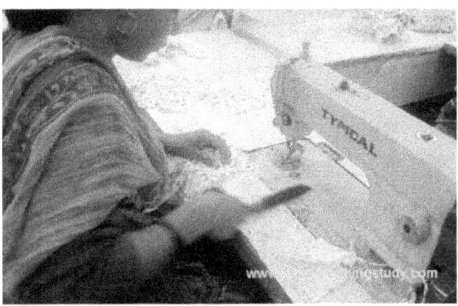

Purpose: Single needle lock stitch machines are used for joining two or multiple fabric plies together. The machine is used to sew lightweight, medium weight and heavy materials.

Fig-1: Lock stitch machine

2. Overlock Sewing Machine

Overlock machines are available in 3 threads, 4 threads and 5 threads over edge sewing. An overlock machine can form various type of stitches like, stitch class 503, stitch class 504 and stitch class 512.

Purpose: This machine is used for serging garment panels (for example: trouser panels serging) and for overedge stitch. These type of machines are mostly used in knitted garment sewing for overedge stitch. Like side seam stitch of a t-shirt is done using an overlock machine.

Fig-2: Overlock machine

3. Flatlock sewing machine

This machine is called a cover-stitch sewing machine. Flatlock sewing machines normally come with 2-3 needles. For bottom cover stitch machine 2 needle threads pass through the material and inter loop with 1 looper thread with the stitch set on the underside of the seam. Flatlock sewing machine form stitches like Stitch class 406.

Flatlock machines are available in two types - Flatbed and Cylinder bed.

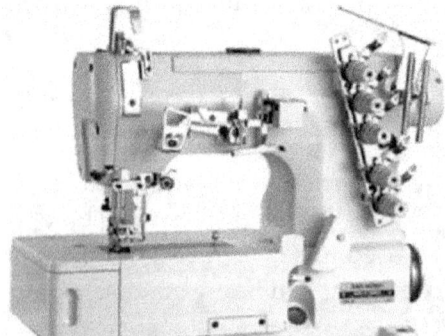

Usage of these machines:
Flatlock machines are used for hemming sleeve and bottom of the knitted products. A cover-stitch machine can be used in any part of the garment for decorative purpose.

4. Feed off the Arm

Fig-3: Flat lock machine

This machine is used in making flat and felt seam. Two needle threads form the chain stitch. For example, this machine is used for sewing shirt side seams and under arms, and for sewing jeans inseam.

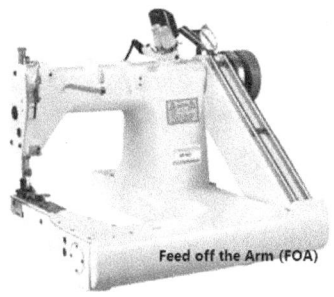
Fig-4: Feed off the Arm machine

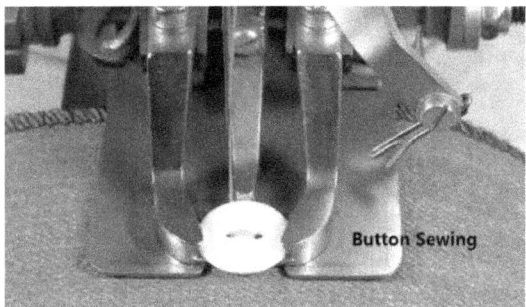
Fig-5: Button sewing machine

5. Button Attaching Machine

A special machine used only for stitching button in a garment. Different sizes of button can be attached using the same the machine by changing the settings.

Purpose: Attach button. Machine stitches buttons and trims thread automatically.

6. Buttonhole Machine

This machine is used for making the buttonholes on garments. Buttonholes can be made with different stitch density. Like in Shirts, Trousers, and Polo Shirts etc.

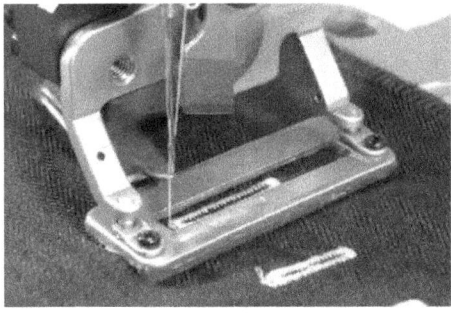
Fig-6: Buttonhole machine

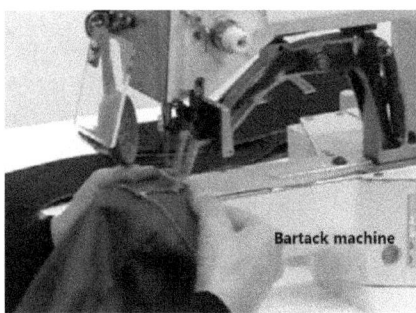
Fig-7: Bartacking machine

7. Bar tack machine

Bar tack stitch is made to reinforce the seam and garment component. Like in belt loop joining and at the bottom of side pocket opening bar taking is done.

8. Zigzag sewing machine:

This machine is used for zigzag stitching. Used in bra manufacturing in attaching elastic tape on the edges, and in jacket manufacturing for attaching inside label.

9. Multi-needle chain stitch machine

This machine used for smocking operations and pin-tuck operations. A four-needle chain stitch machine is used in attaching waistband on a jeans, attached front placket in formal men's shirt.

10. Double needle lock stitch machine

A double needle lock stitch machine is used to sew two stitch lines at a time on the garment part. This reduces stitching time where double stitch line is needed to sew.

For example, attaching pockets in a Cargo pant.

Other than these 10 different types of machines, there are few more types of machines used in making garment.

Different Type of Sewing Line Layouts

What is line layout?

The sewing line layout can be defined as the way sewing workstations are placed in the sewing floor to form a line (or batch) that works on single style. The purpose of choosing a line layout over other is to achieve the highest production utilizing existing resources.

The line layout in a factory is not changed frequently. Line layout is designed at the time of plant set up and after that if factory wants to change production system, they might need to redesign the line layout. You may reallocate sewing machines while setting line for new styles, but you do not change the format of the line.

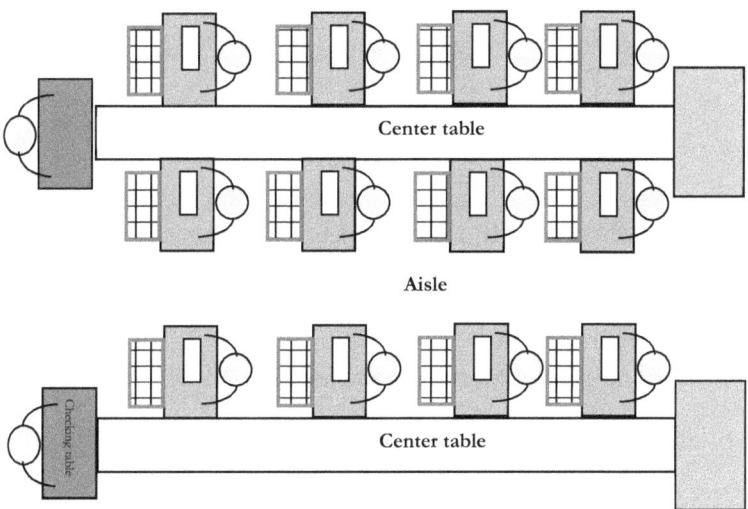

Figure: Line Layout diagram

Different Types of Line Layout

Though there are multiple options of sewing line layout to choose from, most garment manufacturers are comfortable with straight lines having center table in between two rows of machines.

In this article, I will be showing you different types of line layout found in the garment industry. I will not tell you which line layout is good, and which one is not when comparing with others.

The common sewing line layouts are as following.

1. Line with center table and operators facing same direction (Figure-1):

In the line a center table is placed in between two rows of sewing machines. All operators sit on workstation facing same direction. Operators pick bundles from center table and after stitching dispose bundles on the center tables.

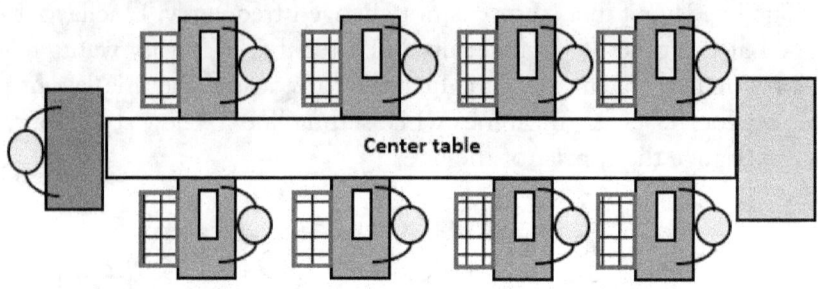

Figure 1: Straight line operators facing same direction

2. Line with center table and operators facing opposite direction (Figure -2)

Machine layout is same as above one. The difference is operators' sitting position. Operators sit on the machine keeping center table left side. This layout is more convenient to all operators for picking up work from left side.

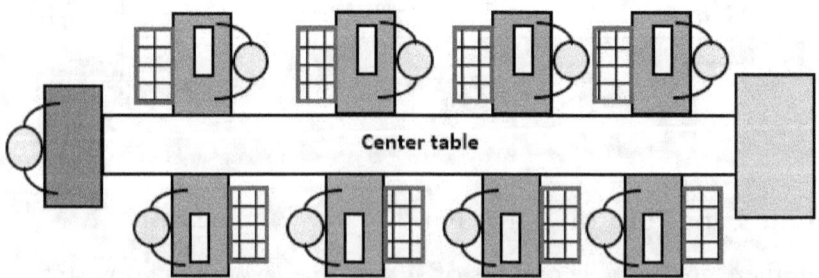

Figure-2. Straight line operators facing opposite direction

3. Straight line without center table and one raw of machines

In this layout no center table is used for material handling. Instead, cutting, and finished garments are kept of hangers, on baskets or on trolleys. See the different form of layout where machines are placed in a straight line, but no center table is used.

Overhead material movement (Fig-3): Garment components are clamped on hangers and transported on a rail.

Fig. 3: Straight line layout with overhead material transportation

Trolley for material transportation (Fig-4): In this line layout, trolleys are used for material transportation instead of installing a center table.

Fig-4: Straight line layout with trolleys

(c) Line having individual disposal basket (Fig-5): Instead of center table individual disposal baskets are provided to operators.

Fig-5: Straight line layout with individual disposal basket

4. Side by side machine layout:
In this layout sewing machines are placed side by side. Two rows of machines are facing each other. This type of layout is used for single piece production system.

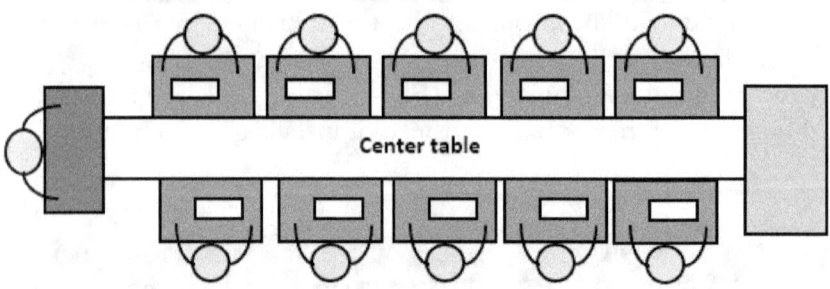

Fig -6: Side by Side machine layout (operators sitting side-by-side)

5. U-shaped line layout:
This kind of line layout is used in lean manufacturing. Machines are placed side by side and U-shape is formed to make a line. Normally, operators sit inside the U shape. No center table is used. This line layout is also known as modular line.

6. Modular line layout:
In lean manufacturing, to reduce material transportation and increase the machine utilization sewing machines are placed in such a way that neither it forms a U-shape nor a straight line. Instead, machines placed that suits

better to work into multiple sewing machines sitting on a single chair. This layout is named as modular layout to differentiate from the above one.

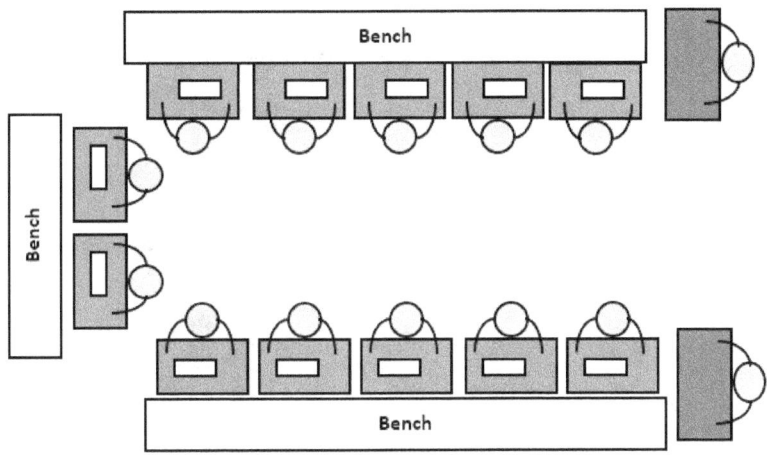

Fig. 7: U-shaped line layout

7. Machine layout in UPS system (Fig-8)

I found this while searching on the web. Machines are placed in straight line but in an angle. In the other UPS workstation machines can be placed side by side.

Fig-8: Line layout in UPS system

8. Unconventional Line Layout (Fig-9)

In this layout sewing machines are placed in zig-zag form and a wide table (same height of sewing table) is placed for material handling. The same table is used for garment checking. See the following figures.

This image of an unconventional production line layout shown to give you an idea for thinking differently.

Figure-9: Unconventional line layout

You may choose one layout for your factory from the above list. When one layout is selected out of above, you need to consider the following fine points
- Space utilization - Space required per workstation compared to another line layout
- Means of material transportation
- Order volume and shipment lead time
- Production systems to be implemented

Different Kind of Material Handling Systems

Material handling system plays an important role in improving material flow and increasing production performance in the readymade garment manufacturing industry. If you ever visited a garment factory, you might have seen various kind of material handling equipment used by them.

At the time of new garment factory set up or improvising production system and factory layout, material handling system is considered for better factory performance and smooth material flow. A right material handling system across the factory departments reduces material transportation time, waiting time at work and reduces production delays.

Define material handling system

The means used for transporting work (like fabrics, cuttings, bundles, finished garments and general items) from one place to another, storing materials and protecting material from damage, are called material handling system. It may be an equipment, device, or procedures. For better material handling, equipment is engineered according to workplace design.

As said above there are different type materials handling systems based on process requirement. Whether it is a new setup or an existing setup, factories have many options for choosing one out of available material handling systems. Common material handling systems found garment manufacturing units are listed here.

1. Material handling equipment for cutting department

Cutting department is responsible for cutting garment components from the fabric rolls. Cutting department needs to handle a lot of fabrics and cutting bundles. So, they need material handling tool and equipment like these.

Trolleys: For transporting fabric from fabric store to cutting section. Forklifts are also used for this purpose.

Baker's trolley: For stacking cuttings and transporting cutting bundles to stitching floor

Plastic trays and trolleys: Cut pieces are placed on treys. Trays filled with garment with certain number of pieces are placed in trolley.

Fabric Bags: Cuttings are bagged into fabric bags or plastic bag and carried out to production section manually.

Racks: Racks used for storing cutting bundles.

2. Material handling equipment for stitching department

Garment stitching department stitches garments from fabric (cuttings). Different types of line layout for stitching line are found in the garment industry. Based on the production line layout, material handling system is designed. In stitching department, material handling system is needed for transporting bundles from one workstation to next workstation, and for storing WIP.

Conventional side table (with bin): A side bench or a side table is placed for loading cuttings and disposing stitched garments. This kind of layout is normally found in group production system and make-through production system.

Centre table: Centre table is the most common material handling system used in the production line (progressive bundle system and single piece production system). A bench (2-1.5 ft wide) is placed in between two rows of machines.

Trolleys: Cutting bundles are stacked on trolley and the trolley is moved in the line. Some factories found it difficult to assess WIP in the line by using trolley system. These trolleys are useful for trouser manufacturing. See image

Plastic crates and Bins: Bins are used for disposing stitched garments. Plastic bins are used for transporting garments from one place to another.

Overhead production system (unit production system): Many production units use overhead rail and hanger for transporting material. Factories found it is easy to track WIP on the UPS system. Some factories experience issue in line balancing. Different types of overhead hanger system are available. Like manual hanger system, semi-automatic hanger system and computerized hanger systems.

3. Material handling equipment for finishing department:

In the finishing department, garments are pressed, checked, folded, and packed. In finishing section, garment pieces are moved to number of workstations before pieces are packed into the polybags.

Common material handling equipment found in the finishing section includes –

Trolleys: Transporting garments from one workstation to another workstation.

Bench: Used for stacking pressed garments

Hangers and stand: Sometimes garments are not folded and shipped in full length in hangers. Pressed garments are hanged on hangers and stored in a stand.

In this article, you saw the commonly used material handling systems and devices in the garment factories. You can customize the material handling equipment based on your need.

Grading of Sewing Operators

Normally, workers in the garment industry, are categorized as skilled, semi-skilled and unskilled worker for deciding wages grade for them. Some garment factories grade operators as A, B or C according to their experience and expertise on the job. If grading of the operators is done scientifically, it will help in selection of operators during line setting of a new style.

Operator grading works as employee motivator. When operators upgraded to the higher grade, they get motivated and give an extra efforts for the next jump.

In most of the factories, operator's salary is decided according to their grade level (when operator is paid on salary basis).

In this article, I will explain how factories do grading of their operators. Consider this one way of doing operator grading. Grading of the operators is done in three steps.

1. Grading of the operations:

First, operations those are performed in making garment are graded based on the following parameters.
1. Critical and non-critical operation,
2. Skill level required to perform an operation,
3. Machine used, such as Single Needle Lock Stitch, Over edge, semi-automatic machine, or Flat Lock machine and
4. Area of operation in the garment.

For an example, list of operations has been shown here for a shirt manufacturing factory. Operations are graded into three categories, such as A, B and C graded operations. Operations classification according to different grades are shown in the following table (this table is made based on factory data)

Table-1: Grading of operations for a shirt factory.

	Grade A operations	Grade B operations	Grade C operations
1	Attach neck band & collar with pattern	Attach yoke & edge stitch yoke	Run stitch neck band
2	Set sleeve placket & stitch arrow	Attach main label to yoke	Run stitch collar (profile)
3	Match & set collar	Hem front placket	Hem cuff or pocket

4	Close collar	Attach fashion pkt	Close flap at bottom
5	T/S arm hole	R/S collar, cuff, flaps	Locking operations
6	Match &set cuff	T/S collar & shoulder	Stay stitch or gathering
7	Close cuff	Join shoulder	Sew pleats
8	Close cuff (folder)	Sleeve attach	O/L small parts
9	Attach front placket	T/S arm hole (SNLS)	Prepare main label & W/C label
10	Hem bottom	R/S side seam	
11	T/S side (FOA)	Hem bottom	
12	T/S side (DNLS)	Hem side slit	
13	Attach pocket	Hem sleeve	
14		Attach side seam	
15		O/L side seam & arm hole	
16		Sew dart (manual)	

You can define grading criteria of the operations as per your standards and include more products in this list.

2. Operator grading criteria:

The factory uses parameters such as how many operations an operator can perform from the operation list shown in above Table-1 and what is their efficiency level at work.

In the Table-2 required criteria are mentioned against each grade. Grading of operators are done as A+, A, B, C, and D.

3. Study of the operator's skill on operations:

To grade an operator, ask the operator to perform operations listed on the above table. See how many operation he/she can do with required efficiency level. Also check their performance level (Efficiency %) on the operations. For the operator's performance study, use a format for data capturing and store all the operator skill for future reference.

Once you have detailed study of an operator, match it with table "operator's grading system" and mark operator. This grades are then added in the skill matrix.

Table-2: Operator grading Criteria

Grades	Performance Criteria
A+	Can perform almost all A grade operations
	5-6 B grade operations
	Almost all C grade operations
	Efficiency >= 56%
A	Minimum 5-6 A grade operation
	5-6 B grade operations
	Almost all C grade operations
	Efficiency >= 52%
B	Minimum 2 A grade operation
	Minimum 4 B grade operations
	Almost all C grade operations
	Efficiency >= 50%
C	Minimum 5 B grade operation
	Almost all C grade operations
	Efficiency >= 45%
D	Trainees
	Any B or C grade operations

Remember, the above grading criteria is followed by one factory. You can follow the same or create your own grading criteria following these steps.

References: The operator grading criteria and method are taken from a shirt making factory based in Bengaluru.

Skill Matrix of Sewing Operators

Skill Matrix is a chart or a database where operator's past performances on various operations are recorded in a systematic way for the future reference. In a skill matrix, operator performance is recorded in efficiency percentage. Skill matrix is also called as skill inventory of the operators. A basic skill matrix is shown in the following Table-1. In this example, operators' individual efficiency (overall efficiency) is listed operation-wise and machine type-wise for a specific product.

Operators' skill matrix is updated on a regular interval by IE team. Or after completion of each style operator's current performance (efficiency%) updated on the database (for manually developed skill matrix).

Table-1: A Basic Skill Matrix developed in a spreadsheet

Emp. code	Emp name	Operation	Operation-1	Operation-2	Operation-3	Operation-4	Operation-5	Operation-6	Operation-7	Operation-8	Operation-9	Operation-10	Operation-11
		Machine Type	SNLS	4THOL	SNLS	4THOL	3TFL	DNLS	4TOL	3TFL	SNLS		
1	Amit		81	76		75							
2	Sumit		46	93									
3	Ram				71								
4	Nitin					80	75	82					
5	Ajay				76								
6	Rani				74			86					
7	Rina								77	56	67		
8	Abhay						77			66	72		
9	Azad												
10	Anil						80						

The benefits of the skill matrix are listed below.
- It keeps record of all operations an operator had done in the past and employee's efficiency level in each operation.
- Industrial engineers and line supervisors need least time to find pick an operator and they can select most efficient operators for an operation from the pool of operators.
- For line balancing, operators can be selected according to work content. For example – where an operation is required 50% less time than pitch time, engineers can select an operator whose efficiency level is 50% on that operation.
- When operation clubbing is required (for less work content works), skill matrix gives the information what all operation to be given to an operator.
- When someone is absent, supervisor can easily find suitable person from the skill matrix table and replace.

The employee skill matrix can be made in a spreadsheet. You can maintain the operator's skill matrix data in an Excel file.

A real-time shop floor data tracking system can be used for skill matrix development. If someone uses real-time data capturing system, skill matrix can be developed automatically.

In the skill matrix, a user can do sorting of operator performance in various operation and other way in an operation how many operators are there who had earlier worked on and their efficiency level.

In the next article, I will show you how to develop operator's skill matrix in a spreadsheet.

How to Develop Skill Matrix? (using an Excel Template)

In this article, I will show you how to prepare skill matrix template in a spreadsheet and quickly retrieve the skill history data when you need it. You might already know what the skill matrix is. It has been explained in the previous article.

Objectives of developing and maintaining operator skill matrix

Before you start developing database for operators' skill matrix, you should be clear about the objective of developing skill matrix database. If you ask me what the object of developing skill matrix would be, I will tell you primary objectives are -
1. To check operators' skills in handling different machines (in efficiency percentage).
2. To check operators' capability of doing various operations (If you maintain operations wise record, you may skip recording machine record or keep both) and their performance level (efficiency %).
3. To identify operators who can do the selected operations and measure their performance level (efficiency).

Method of developing Skill Matrix

Step by step guide to develop template for skill matrix.

Step#1: Develop excel template for data recording:

We will be using Excel sheet for developing skill matrix. (Reason, most of the garment factories do not have software for recording and storing operators' skill matrix data.)

To have all basic information in one table (Spreadsheet), you need to capture following data.
- Date of data entry,
- Operator name and employee code,
- Operation name,
- Machine name and
- Operator efficiency on the operation

Minimum information you need to capture is shown in the following table (Table-1)

Table-1: Skill matrix data entry sheet

Date	Operator Name	Employee Code	Operation Name	Machine Type	Efficiency
15-Feb	Sagar K	1120	Back yoke attach	SNLS	56
15-Feb	Sunita R	1230	Shoulder attach	SNLS	70
15-Feb	Anita	1233	Shoulder T/S	SNLS	52
15-Feb	Latha	1125	Sleeve Hem	SNLS	52
20-Feb	Kishan	1121	Pocket Hem	SNLS	65
15-Feb	Kishore	1132	Pocket attach	SNLS	45
1-Mar	Rajaram	1221	Collar making	SNLS	87
1-Mar	Kuldeep	1234	Sleeve hem	SNLS	44

Sometimes, you may be interested to have detailed analysis about operators' performance, production quantity, style number on which operator has worked, number of hours she has worked etc. Capturing that much data manually and entering the data into computer is time consuming. Just for information one more data entry table is shown here (Table-2).

Table-2: Skill matrix data entry sheet

Date	Operator Name	Emp Code	Style	Operation Name	Machine Type	Prod. Qty	Shift Hrs.	Eff%
15-Feb	Sagar K	1120	54120	Back yoke attach	SNLS	400	8	56
15-Feb	Sunita R	1230	54120	Shoulder attach	SNLS	450	8	70
15-Feb	Anita	1233	54120	Shoulder T/S	SNLS	400	8	52
15-Feb	Latha	1125	54120	Sleeve Hem	SNLS	400	8	52
20-Feb	Kishan	1121	54120	Pocket Hem	SNLS	542	8	65
15-Feb	Kishore	1132	54120	Pocket attach	SNLS	550	8	45
1-Mar	Rajaram	1221	54120	Collar making	SNLS	400	8	87
1-Mar	Kuldeep	1234	54120	Sleeve hem	SNLS	433	8	44

Step#2. Capture data:
Collect employee wise production information as shown in the Table-1. For individual operator efficiency data calculate it using this formula.

> Operator efficiency = (Production quantity × Operation SAM /Total minutes worked) × 100

You may capture all records of an operators but maintaining the Excel-based skill matrix, you do not need to enter everything in the database. You will get lot of variation in her efficiency on different dates and different styles. This happens due to many reasons like having lost time, line setting, no feeding, working for few hours in a new operation, working on new style etc. Ignore abnormal data and only record best performance out of many different efficiency data against an operator.

Step#3. Enter data in the excel template

Instead of updating this table daily, you can update weekly interval or update when you see an operator doing a new operation or having good performance. (Though operators' performance depends on various parameters).

If an operator is doing multiple operations in a same day, enter their data in different rows (operations-wise entry) to have her performance in all operations she has done.

You can keep old record or replace old one with new record for an operator doing same operation.

Step#4. View skill history report as you needed

I have made pivot table to fetch data from the table (database) where you will be entering data.

In the following section, I have explained how to use excel templates for data entry and reviewing reports.

How to use Excel Template to retrieve skill history of the employee/ operations?

The objective of making database is to get information on
- Employee performance analysis
- Selecting right employee for selected operations when required (also at the time of line setting)

I will show you how to use excel template with an example. I have made two sheets – data entry sheet and data retrieve sheet. I guess you can operate excel sheet very well. See Table-1 for data entry template.

1. Data entry sheet

In this sheet, you will enter employee data with all required information. As said above you can modify earlier data of an employee by replacing it with new record. Or add a new record.

2. Refresh file

When you add new record to the data entry sheet, refresh the file (Click menu Data - refresh all). This is done to get updated data in the second sheet. See below screen (Figure-1)

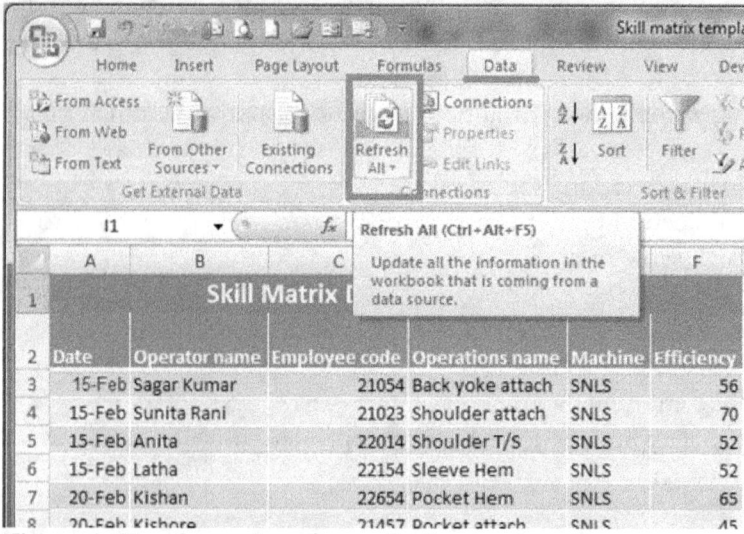

Figure-1: Excel template data refresh

3. Data retrieve/ Review reports

In the data retrieve sheet, I have added two tables named
- Operator wise data view
- Operation wise data view

Operator wise data: Select operator number for the dropdown menu (See figure 2). You get list of operations done by that selected operator, list of machine type she had operated before and date when you have last updated her record.

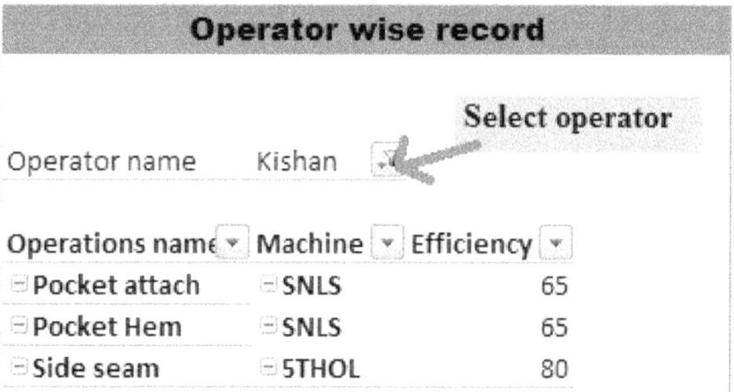

Operator wise record

Operator name	Kishan		
Operations name	Machine	Efficiency	
Pocket attach	SNLS	65	
Pocket Hem	SNLS	65	
Side seam	5THOL	80	

Figure-2: Operator wise skill history

Operation wise data: Select operation name for the dropdown menu (See Figure. 3). You will get list of employees who all did the selected operation (in the example operation: Sleeve hem), list of machine type she had operated before and date when you have last updated her record.

Opration wise record

Operation name: Sleeve Hem

Employee code	Operator name	Machin	Efficier	Date
20135	Kuldeep	SNLS	44	3/1/2016
22154	Latha	SNLS	52	2/15/2016
20014	Shyamal	Flatlock	65	3/5/2016

Select operation from the list

Figure-3: Operation wise skill history

Developing and maintaining skill matrix is a time-consuming task. Still if you develop it once and update the same in weekly or monthly interval would be very helpful. The real-time shop floor control systems have such feature to record operator skill history automatically.

If you have good knowledge in Excel and pivot table, you can develop this type of skill matrix by your own.

Garment Production Cost: Actual Cost Vs Cost Per SAM

One of the most important production KPIs in garment manufacturing is the comparison between actual garment production costs and cost per SAM.

Factories use to calculate actual production cost and cost per SAM, and they compare data on daily basis. The actual cost figure shows how much money (wages) factory is paying to the operators. The different methods of calculating actual production cost and garment cost per SAM are explained here.

Actual Production Cost per Garment

The total cost incurred in making garments in the production floor is known as the actual production cost. It includes salary/wages of all the direct and indirect employees involved in producing garments. Factory overheads are not included here. Production a calculated on daily basis.

1. First method of calculating actual production cost:

Formula of the actual labor cost of a garment-

> Actual production cost = (Total salary to be paid/Total Units Produced).

For a line, calculate total salary (W) to be paid on the day including direct manpower. Find total units produced (U) in shift time (output of the line).

2. Second method of calculating actual production cost:

> Actual production cost = (Total salary to be paid/ (Total Minutes produced/Garment SAM)).

Though the first method is used by most of the factories, it is not true actual cost. Because, only output quantity is considered in this cost calculation, where a lot of works is produced on the line as WIP. There are two main reasons why engineers prefer the first method.

- It is very difficult to find how many units have been produced by individuals in a line and then calculating total minutes produced in a day.
- To the factory management, it does not matter how much work is laid (partially stitched) in the line. Total garments come out from the line are considered as production.

I will also suggest you use the first method as because of it simple to calculate and easy to understand. This cost goes down day by day during the learning curve. Secondly, every day you will get different garment production cost based on the line output.

Cost per SAM

The formula used to calculate the cost per SAM.

Cost per SAM = (Garment SAM × Cost Factor).

Here the cost factor is per minute labor cost to the factory at factory average efficiency. Simply instead of cost factor you can multiply Garment SAM by average labor salary per minute (which represent cost factor calculated at 100% efficiency). Whatever method you use, it must be clear to everyone (persons who see the report) within the factory. Cost per SAM is fixed for a style.

Cost factor = Labor wages per day / (shift minute * Efficiency).

Let's say, shift time 480 minutes, daily operator wages ₹200.00 and factory run at 50% efficiency. Then cost factor for the factory will be 0.833.

Example: Factory's efficiency 50%, Daily wages ₹200.00 Shift time 480 minutes. See the comparison in the following table where style A and style B's Cost per SAM and actual cost are shown. Use above formula to calculate figures.

Style	SAM	Manpower	Planned Production	Actual Production	Cost factor	Cost /SAM (₹)	Actual Cost (₹)
A	15	20	320	150	0.833	12.5	26.67
B	12	20	400	250	0.833	10	16.00

In the next article, you will learn the method of calculating actual garment production cost.

103

The Way Factories Calculate Production Cost

In the daily production report, many factories include the actual garment production cost from the day of loading a style. In this article, production cost represents sewing room cost. Factories consider all direct and indirect personnel who are involved in garment sewing and managing sewing lines in calculating garment production cost.

In case, a production manager handles 10 lines then one-tenth of his daily salary is considered as cost incurred per line per day. If a floor-in-charge handles 3 production lines, (considering each line run single style) then one-third of his daily salary is considered as cost incurred in the production of a style.

By following the above-mentioned cost sharing method, daily salary of all employees is calculated for a style. If operators work overtime and others stay to assist them then overtime salary is also added to total daily wages. Overtime salary (hourly pay rate) may differ from the normal working hours.

"Actual Garment production cost" is a component of daily production report (DPR). Actual garment cost is represented in two ways –

- Actual garment production cost per unit for the day and
- Average production cost per unit (calculated from the style loading date to till date)

The formula used for the cost calculation

> Actual production cost = (Total salary to be paid/Total Units Produced).

In the following table, an example of daily production cost calculation has been shown that is normally used by garment manufacturing factories.

In the below example, production manager handles 5 sewing lines and Floor-in-charge handles 4 lines. So, their salary is divided by 5 and 4 respectively as the expenses share to each line. If you want to use this sheet for your factory, add manpower in the list and their actual daily wages (salary per day) according to your factory.

Daily Production Cost Sheet

Line No.: 5A
Style: TT12977
Production day no.: 11
Date: 03/04/2011
Output: 250 units

Sl. No.	Employee Designation	Number of employee used (A)	Daily rate (B)	Today salary (C=AxB)	OT rate (E)	OT hours (F)	OT amount (G=ExF)	Net salary (H=C+G)
1	General manager	0	4000	0.00				0.00
2	Production manager	0.2	1000	200.00				200.00
3	Floor-in-charge	0.25	600	150.00				150.00
4	Line supervisor	1	410	410.00				410.00
5	Q.A. manager	0	410	0.00				0.00
6	Quality checkers	2	200	400.00	50	2	100	500.00
7	Pressman	0	200	0.00	50			0.00
8	Handworker	0	215	0.00	50			0.00
9	Machine mechanic	0	180	0.00				0.00
10	Helpers	1	172	172.00	40	2	80	252.00
11	Sewing operators	15	200	3000.00	50	30	1500	4500.00
12	production writer	1	200	200.00	50	2	100	300.00
Total		20.45		4532.00	290.00	36.00	1780.00	6312.00

Previous Cumulative wages (P)	Today's Wages (Q=H)	Total wages till date (R=P+Q)	Previous Cumulative production (S)	Today's output (T)	Total output till date (U)	Average cost till previous days (V=P/S)	Today Cost (W=Q/T)	Average cost till date (X=R/U)
61000	6312.00	67312.00	2100	250	2350	29.05	25.25	28.64

(Note: All salaries and manpower used in the above costing sheet are only a representation.)

During the learning curve, production cost per piece will be high, and as learning curve ends daily production cost goes down. At this time, the actual production cost is being checked with target cost. It helps a company to assess factory performance in term of meeting garment cost.

On-Standard Efficiency and Overall Efficiency

When it comes for measuring production line efficiency, the efficiency is measured in various forms. Standard Efficiency, off-standard efficiency, Overall efficiency, and True efficiency.

In this article, I will explain on-standard efficiency and overall efficiency. Then we will see the difference between on-standard efficiency and overall efficiency

The efficiency of a production line and an individual operator is calculated by using following formula.

> Efficiency (%) = (Total minutes produced / Total minutes attended at work) x 100

What is Overall Efficiency?

The above formula gives us the overall efficiency of an operator or a line. In the above example, all working hours are considered in the calculating efficiency. In the total shift hours, sometimes operators will work a standard job, sometimes they will work on off-standard jobs and sometimes they may not be doing any job due to non-availability of work (i.e., no work). When the total worked hours are not separated as on-standard worked hours and off-standard worked hours, and total hours are used in the calculation, it gives us overall efficiency.

Let me explain this with an example. In the example, I have taken two operators. They worked for 480 minutes in a day. Out of 480 minutes they spent time on the standard-work, and they had some loss-time (minutes) as shown in the Table-1.

Table-1: Overall efficiency calculation

Seq. no.	Employee code	Minutes produced	Standard hours worked	Loss time	Total hours worked	Overall efficiency
	A	B	C	D	E=(C+D)	F= (B/(E x 60)) x 100
1	20031	360	430	50	480	(360/480) x 100 =75 %
2	20034	300	410	70	480	(300/480) x 100 =62.50 %

106

What is On-Standard Efficiency?

When the efficiency is calculated using on-standard produced minutes and on-standard attended hours, it is known as on-standard efficiency.

In a shift, operators attend 480 minutes (8 hours shift), but they may not always sew garments. Sometimes operators wait for work due to external reasons or they do off-standard jobs (operations) in the 480 minutes duration. Reasons are like waiting for work, machine break-down, power failure, line setting, quality related issues, non-availability of trims and accessories. All these off-standards reduce operator's real performance (efficiency level).

When operators are not working the **standard jobs,** they are simply not producing any garments and not producing any standard minutes. That is why, to know the operator's actual performance on the standard jobs, the operator's efficiency is presented as On-standard efficiency.

The formula for calculating on-standard efficiency is same but attended time. In case, of overall efficiency calculation total produced minutes is divided by total attended minutes at work. But for the On-Standard efficiency calculation, total produced minutes is divided by total attended minutes at standard jobs.

The On-standard efficiency calculation formula:

> On-standard Efficiency (%) = (Total standard minutes produced / Total minutes attended on on-standard operations) x 100

Taking the same example, here we will see on-standard efficiency calculation for those two employees. Here we will consider only on-standard hours worked time. See Table-2.

Table-2: On-Standard efficiency calculation

Seq. no.	Employee code	Minutes produced	Standard hours worked	Loss time	Total hours worked	On-standard efficiency
	A	B	C	D	E=(C+D)	F= (B/Cx60) x 100
1	20031	360	430	50	480	(360/430) x 100 =83.72 %
2	20034	300	410	70	480	(300/410) x 100 =73.17 %

Let's compare overall efficiency and on-standard efficiency of these two employees and data is shown in the following Table-3.

Table-3: Overall efficiency and On-standard Efficiency

Seq. no.	Employee code	Overall efficiency	On-standard efficiency
1	20031	75 %	83.72 %
2	20034	62.50 %	73.17 %

Difference between On-Standard Efficiency and Overall Efficiency

The on-standard efficiency of a line or an operator will be higher than or equal to the Overall Efficiency. If the employee works on the standard job for the whole day, the On-standard efficiency and Overall Efficiency value will be equal.

In case, an operator does an off-standard job (spent sometimes in the off-standard job, non-productive times), the On-standard efficiency will be higher than the overall efficiency.

Calculating the overall efficiency (for a line or an operator) is easier compared to calculating the On-standard efficiency, as to calculate On-standard efficiency, you need to measure time spent on the standard operations and you need to measure the non-productive time.

Major Factors Affecting Factory Efficiency

What are the major factors that affect factory efficiency directly?

The right answer to this question depends on the current factory practices. Depends on the production systems, workflow, and effectiveness of the managerial team. The direct reason may vary from one company to another.

In general, the following are the key factors that may cause maintaining and improving factory efficiency in garment manufacturing.

1. An imbalance production line

If you are having assembly lines and workload are not balanced in a line between workers, some workers will not get enough work for the day. And a few workers will have a pile of works. This situation causes underutilization of standard hours which in turn affect the line efficiency.

2. Unmeasured lost time and no action on lost time reduction

The time when workers are not doing standard work and waste their time being idle is called as lost time. Many times, in a day, operator wait for work (feeding). Sometimes they wait for trims. Operators wait for the line to set up. Factory loses productive time that affects the factory's efficiency.

3. Unskilled employees

The production rate of a highly skilled worker is higher. An unskilled operator works at a low production rate compared to skilled employees. Secondly, unskilled operators make mistakes and produce defective pieces.

4. Not using time-saving equipment, folders, and attachment

This kind of time-saving tools can reduce operation cycle time and can help to improve the operator's efficiency.

5. Demotivated workforce

Production speed in the shop floor in a factory also depend on the workers' mood. Demotivated workers can reduce the factory's performance. Untimely payment and low wages may demotivate workers.

6. Stitching quality issues

When quality issue is there, the operator needs more explanation and training, which in turn reduce the operator's production rate, and it cause efficiency fall of the given line. Secondly, if the garment quality is not made right the first time, there will be repair work, rejection, and requirement of additional manpower.

7. Wasting productive time during style changeover

Excess waiting time for line set-up during style changeover cause factory's efficiency fall. This mostly happens when line supervisors, production team and engineers are not pro-active and not ready with manpower and machine allocation, line layout, work aids, and pilot run.

8. Failing to monitor production

No monitoring of hourly production performance and not chasing the operators for target quantity and efficiency, cause efficiency fall.

9. Employee absenteeism

When some of the employees do not come to work, it causes production loss, creates an issue in the workflow. If the workers leave or absenteeism is not planned, it can reduce line and factory efficiency.

10. Machine breakdown and maintenance delay

Machine breakdown is a common fact. It may happen. But if the machine maintenance team takes a longer time to resolve the machine issue, line efficiency will fall.

Conclusion

The factory efficiency is one of the many business performances measuring tool in the manufacturing sector. Factory's growth and earning is directly linked with factory efficiency. That is why every business owners must take care of improving and maintaining factory efficiency.

You learned about the factors (root causes) for the low factory efficiency. Through improvement project, you can reduce the impact of these factors and increase factory efficiency.

Difference between Productivity and Efficiency

Productivity and efficiency are the primary performance measures in garment production. We have earlier discussed the Productivity and Efficiency terms and calculation methods separately. Here I will explain the difference between productivity and efficiency with an example.

First let me define these two terms.

Productivity is the ratio of output and input (labor, machine, man-hours). Productivity can be measured as labor productivity, and machine productivity for a production line or of a whole factory. Let us consider you are measuring labor productivity of a production line. Labor productivity is defined as output (in pieces) per labor (direct +indirect) in the given time frame. An example of labor productivity of a shirt line is 10 pieces per day per labor in 8 hours shift.

Efficiency is the ratio of total minutes produced and total minutes worked by an individual operator, or a line. Efficiency is expressed in percentage. For example, stitching line efficiency of a typical stitching line is 56%.

Formula used to calculate productivity and efficiency

Productivity formula:

Productivity = Output / Input

Labor productivity = (Total garments produced / Number of labors involved) in the given time frame

Machine productivity = (Total garments produced / total number of machine utilized)

Efficiency Formula:

Efficiency = (Total SAH produced / Total hours worked) x 100
Efficiency is measured for individual operators, for each lines and overall factory efficiency.

In garment production, labor productivity is presented as number of pieces produced per labor per shift.

Labor productivity of the same line with same number of labor for two different products (e.g., a T-shirt and a Polo shirt) may be different. But Line efficiency can be same in both products.

Productivity is product specific measure. When you compare your factory's productivity with another factory's productivity, you need to check if both factories are making same product. If different products are made by the factories, you can do productivity comparison by converting productivity of one product to equivalent SAM of the second product.

To calculate line efficiency, you must have garment SAM (standard minutes) of your styles. On the other hand, you do not need SAM to calculate labor productivity. Factories those do not have industrial engineering department and do not calculate garment SAM are not in position to measure accurate line efficiency. For them productivity measure is one that they can follow.

Both have importance in understanding factory's performance. In case a factory has multiple stitching lines, and all lines are producing same products/design without any variation in construction and all line has same number of manpower, in such case labor productivity measure is enough to compare performance between the two lines.

But how will you compare if these two lines are making two different products? And have different number of manpower. With productivity data you could not do that. In such situation, line efficiency is the only measure to compare production performance of these two lines.

Let us take one case for calculating labor productivity and line efficiency. A line of 35 operators and 5 helpers produced 400 pieces in a day of 8 hours shift. Garment SAM is 25 minutes. Considering helpers SAM is included in garment SAM.

Total labor input (A) = 35+5 = 40 labors
Total production (B) = 400 pieces of garments
Total SAM produced (C) = (400 pieces x 25 minutes) = 10000 minutes
Total Minutes attended by all workers (D) = (40 workers x 8 hours shift x 60) = (40 x 8 x 60) minutes
Labor productivity = B/A = 400/40 = 10 pieces per labor per shift
Line Efficiency = (C/D) x 100 = (10,000 x 100) / (40 x 8 x 60) = (1000,000 /19,200) % = 52.08%

Comparison is shown in the following table.

Table: Productivity Vs Line Efficiency

Date	Line Number	Labor Productivity	Line Efficiency
Aug-4	Line-1	10	52.08%
Aug-4	Line-2	8	62.40%

Pitch Time and Pitch Diagram

What is pitch time?

Pitch time is a ratio of total SAM of garment and number of operations to be set for the style.

> Pitch Time = Garment SAM/No. of operations.

Let say a polo shirt SAM is 16 minutes and there are 20 operations in the polo shirt. You want to set-up a line for this polo shirt in 20 workstations. In this case, pitch time would be (16/20) = 0.80 minutes.

Pitch time is used for line setting and calculating production target for a sewing line.

Pitch diagram:

A graphical presentation of individual operation's time (SAM) and pitch time on a same chart is called pitch diagram. For example, see the following chart (Fig.1). At this chart on X-axis operations name and on Y-axis time value is depicted.

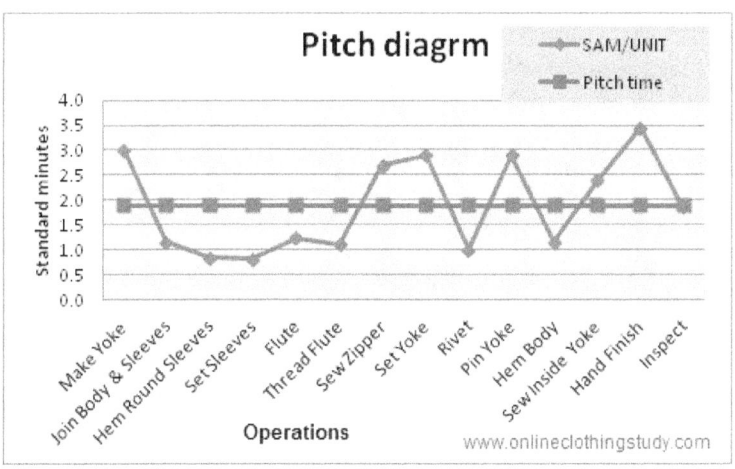

Figure: Pitch Diagram (SAM Vs Pitch Time)

Usage of Pitch Diagram:

- Pitch time is used to calculate machine requirement in each operation.
- Pitch diagram is used for line balancing in an assembly line. Pitch diagram is made on operator's production capacity per hour and target quantity per hour in pieces for easy understanding.

How to make a Pitch Diagram?

To make a pitch diagram on operator's capacity Vs line target collect the following information. First conduct a capacity study for all operators and find out how many pieces operators are making at each operation. Where more than one operator is doing same operation, sum up their capacity for that operation. With the capacity study data make one table on the spreadsheet as shown in the following the Table-1.

Table-1: Operation breakdown on a style (data shown in the following table and chart are assumed only for example.)

A	B	C	D
Sl. No.	Operations	Capacity/Hour (Pieces)	Target/hr. (Pieces)
1	Collar run stitch	35	27
2	Back dart making	27	27
3	Sleeve slit binder	29	27
4	Front placket attaching	21	27
5	Front placket finish	20	27
6	Front dart making	33	27
7	Shoulder Join O/L	30	27
8	Shoulder edge stitch	40	27
9	Sleeve attach O/L	50	27
10	Side attach	20	27
11	Cuff attach	17	27
12	Cuff finish & Tacking	16	27
13	Collar attach	24	27
14	Collar finish	24	27
15	Bottom Hem	27	27

For example, the line is making ladies blouse. Operations are listed in column B as per operation sequence (all operations are not taken). In column C, number of pieces is listed according to the operations that can be produced by the operators. Add one more column 'D' and write target quantity (in the example -27) against each operation.

Suppose that hourly production target is 27 pieces for the line. So, to meet the target at the end of the line you should get minimum 27 pieces per hour from each operation and for the preparatory operations capacity should be more than 27 pieces. Now draw a line chart with capacity and line target (Fig. 2). This chart is also called as "Pitch Diagram". In the Figure-2, the straight line is presenting the target quantity and second line is showing actual capacity at each operation.

In the chart, the actual capacity line moves up and down. Where the hourly capacity is below the target line, line will produce less quantity than the target quantity. These operations are the potential bottleneck for the line.

On the other way, where hourly capacities are above the target line, it means operators have potential to produce more pieces on those operations if they are provided work. From the pitch diagram, it is easy to understand which operations are bottleneck and where operator's full capacity is not being utilized.

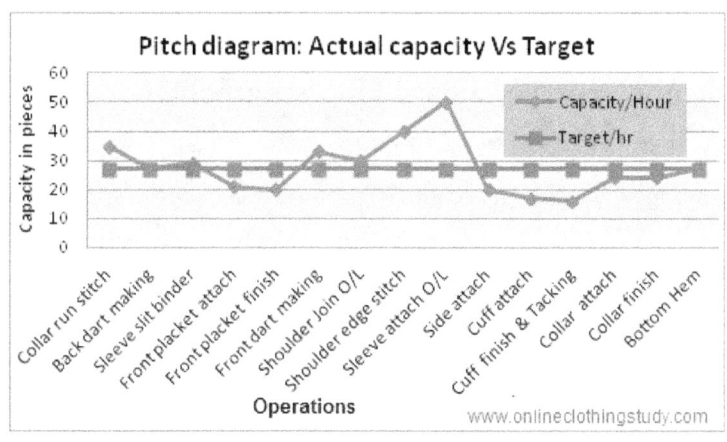

Figure- 2: Pitch Diagram (Actual capacity Vs Target quantity)

Follow the above steps and make your own "Pitch diagram" with real data of a production line.

What is Standard Hour Earned (SAH)?

SAH stands for standard allowed hours. It is a unit of measure like the Standard Minute Produced by workers. Instead of minutes, total produced minutes are presented in hours. The standard hours earned by operators are represented in SAH earned. To get standard hours earned value, total produced minutes are divided by 60 (60 minutes). Like, if an operator works for 8 hours a day and produces garment equivalent to 360 minutes, then the operator's standard hours earned would be 6 hours (This is derived from 360 minutes/60).

The formula used in calculating standard hours earned (SAH) for an operator is shown in the following box:

> Standard Hours Earned = (Operation SAM x Number of garments produced)/60

This term is also called as Earned Hours, because the operator has earned that many hours through making number of garments.

The benefit of using Standard Hours Earned

By measuring SAH you can -
- Compare produced hours against available hours in a day easily (efficiency).
- Secondly, calculating the earning amount (in Dollar) of an operator from earned hours is easy as you know the standard hourly rate of your operators and operations.

Earned hours calculation method:

To calculate the earned hours of your workers, follow these steps
- Estimate SAM for all operations (sewing and non-sewing operations)
- Count garments produced by the operator. You should record operation wise production quantity, as an operator may work in multiple operations, and of different SAM.

- Now follow the above formula to calculate earned hours for all operators one by one. You can use the spreadsheet for calculating earned hours quickly.
- In case one operator is doing more than one operations, first calculate operation wise earned hours, then sum up the operation wise earned hours to calculate total earned hours by an operator.

Let us assume that an employee doing side seam operation. She has produced 300 pieces of garments on that operation. The SAM (standard minute) of that operation is 0.80 minutes. Total minutes produced by the employee is 300 x 0.80 minutes.

Earned hours = (300 x 0.80)/60 hours = 4 hours

For your reference, I have shown few examples of standard hours earned in the following table (Table-1).

Table-1: Earned SAH calculation

Employee code	Pieces produced	Operation SAM	Standard Hours earned (SAH Earned)
2330	410	0.80	5.47
2342	450	0.75	5.63
2354	390	1.10	7.15

Exercise:

In the following table employee wise produced quantity and operation SAM is shown. Operator Bhoomi has done two operations. Calculate SAH earned by these three operators.

Table-2: Calculate Earned SAH and fill the blank cells

Employee Name	Operation Name	Pieces produced	Operation SAM	Standard Hours earned
Dev	Sleeve attach	450	0.80	
Adil	Bottom Hem	400	0.90	
Bhoomi	Sew back pleats	500	0.30	
	Attach yoke to back	500	0.50	

Estimation of Cutting SAM

Cutting SAM can be established in the same way you calculate stitching operation SAM of a garment. In cutting SAM calculation, consider cutting room sub-processes such as fabric spreading, marker making, layer cutting, fusing, re-laying, re-cutting, sorting, layer numbering and bundling as operations. Whether you like to determine only cutting SAM or all the cutting room processes depend on you. I have explained procedures for calculating SAM for all cutting processes.

The important thing, in the cutting process determining SAM of a single piece has no meaning when a cutter cuts parts of several garments at a time. Whether a cutter cuts 10 plies or 100 plies (using a straight knife machine), cutting time will be remain same. Primary variables in determining Cutting SAM are numbers of fabric layers in a lay, number of markers in a lay.

To determine standard minutes for cutting jobs, I suggest you conducting time study of cutting operations, instead of using MTM2 database (synthetic data).

Determine SAM of each cutting tasks using the method explained in the article: *How to calculate SAM of a garment?* Follow these steps

1. Do cycle time of Jobs (time study for 5 – 10 cycles)
2. Do performance rating (Assess how fast or slow the operator doing the job in 1-100 scale)
3. Calculate Basic time of each cutting processes
   (Basic time = Cycle time **x** Performance rating)
4. Add allowances (Machine allowance + personal fatigue)
5. Calculate standard minutes (SAM = Basic time + Allowance percentages)

Spreading SAM

Conduct time study and measure cycle time for fabric spreading process. Convert the cycle time into standard time. Spreading time will vary depending on layer length (and/or number of markers) and type of fabric. To be specific record spreading SAM based on lay length and fabric type. Also record number layer-man involved in layering. Once you have standard minutes for spreading of a layer, you can easily calculate total time required for spreading a lay by multiplying number of layers in a lay. Refer to the following table to record spreading SAM.

Table 1: Spreading SAM record sheet

Sq. no.	Style No.	Fabric type	Lay length (meters)	Marker size	Number of layer man	SAM/layer
#1						
#2						
#3						
#4						

Marker making SAM

Maker making required few minutes and compared to other cutting processes this time is negligible. In case you need to determine marker making SAM and want to add in total cutting SAM, I am showing you how to do it. If CAD marker is used, then do cycle time of lying of the papermaker on the lay and fixing the maker with gum tape.

For manual marking determine standard time through time study. In manual marker, marking time varies depending on garment components, garment size and the number of markers. Record the number of persons involved in marker making. Refer to the following table to record marker making SAM values.

Table 2: Marker making SAM record sheet

Sq. no.	Style No.	Garment size	Marker size	Total components / size	SAM/lay	No. of marker man
#1						
#2						
#3						
#4						

Cutting SAM

Conduct time study for layer cutting operation. Total cutting time of a lay will depend on garment components, linear length to be cut for garment patterns, marker size, and type of fabric. Cutting time also depends on cutting equipment. Record time study data in the following table. Study cutting time for all sizes in a lay and find average SAM. Cover all sizes in time study. Create a database of cutting SAM with the different combination of cutting for future use.

Table 2: Cutting SAM record sheet

Sq. no.	Style No.	Fabric type	Marker size	Total components / size	Lay length (meters)	SAM per Lay	SAM per Garment
#1							
#2							
#3							
#4							

Fusing SAM

Set temperature and pressure of the fusing machine for a certain speed of fusing belt. You can do here reverse calculation. Instead of time study record operator speed for placing cut components (to be fused) in a minute. Once you get production per minute you can calculate fusing process SAM per garment (based on number of fused components) by using formula 1/no. of parts fused per minutes.

Re-cutting SAM

You may need to do re-cutting for some components. Re-cutting is done by other equipment such as hand scissors or band knife machine than straight knife cutting machine. Follow same procedures of determining cutting SAM for re-cutting SAM. For re-cutting process, SAM can be presented per lay or per garment. Also, consider whether all components need to re-cut or only a few components.

For the other cutting room processes like, Sorting, ticketing, and bundling determine SAM as per unit as mentioned in the following table.

Table 4: Units for presenting SAM of cutting processes

	Cutting Operations	Units for cutting operation SAM
1	Spreading	Calculate SAM per layer.

2	Marker making	Manual marking: Calculate SAM for each marker. CAD marker: Calculate SAM per lay (Cut)
3	Cutting	Calculate SAM per marker.
4	Fusing	SAM per garment
5	Re-cutting	Re-cutting SAM can be calculated based on the number of layers cutter cut.
6	Sorting	Calculate SAM per marker
7	Numbering/ ticketing	Numbering SAM can be calculated per garment. For that first, find SAM per component and later multiply how many components are numbered
8	Bundling	Calculate Bundling SAM per marker

In the above tables, you collect SAM of all cutting processes in different units. All these standard minutes will help you to determine cutting room capacity, or you can calculate how much capacity (standard minutes) you will need for an order.

Initially, you need to collect standard time for all cuts to create a database. Once you have created a database with the different combination of lay length, marker size, garment sizes you can use SAM from your database for future styles.

Industrial Engineer's Digest

Role of the Garment SAM in Production Planning

To some extent industrial engineers do the production planning activities. In many garment factories, industrial engineering and production planning department is headed by one person.

SAM of a garment is defined as how much time it would take to complete a garment in sewing. This is also known as garment work content and standard time for making a garment.

To know the role of the garment SAM in production planning, first, you need to understand the primary roles of a production planning and control (PPC) department.

To be specific, in this section, I will point only key roles of PPC department, those cannot be performed without having garment SAM.

- Determining the capacity of the factory and capacity of the individual sewing lines in terms of how many pieces (product specific) factory can make in a certain period with existing machines capacity.
- Order booking based on factory capacity for different types of products
- Allocating of styles to the lines
- Determining production lead time for each order (styles)
- Process scheduling
- Production execution and monitoring

Role of SAM in production planning includes

1. Line capacity calculation: The scientific method of calculating the production capacity of a line (in production pieces per day) is to use standard time (SAM) of a garment. So, to determine the production capacity of a line (for specific products) in pieces you need to know garment SAM.

2. Production lead time calculation: Based on the production capacity, order allocation is done for different lines. A planning guy also needs to calculate how long a style would run in a line if loaded in a single line. If you need to complete the order in less time, calculate how many lines to be considered for an order.

3. Order booking: During order booking, you need to consider capacity availability in a certain period. In such cases, you can use how many minutes you need to make the new orders using garment SAM value and compare the same with how many production minutes are available in your factory for the defined period.

4. Process scheduling: Time and action calendar and production process scheduling of each order is done by the planning department. Again, to schedule a list of tasks, you need to know the capacity of each process per day (or a predefined period). Based on the capacity of each process you allocate no. of days for the process. Like for the sewing department, you determine the sewing capacity of your line (or multiple lines) and according to that, you set how many days to be given to the sewing department for production.

5. Order execution and production monitoring: Standard minutes help planners to set a target for sewing lines. Mutually agreed and calculated target given to line supervisors. Daily when you check production status you can compare actual production with target production. In case production is getting delayed you can push the production team based on the given target.

6. Labor cost estimation: One most important task is the labor cost estimation of a specific order. To estimate how much labor cost to be considered for an order (style), you cannot make labor costing without having garment SAM.

All the above six points are the proof that garment SAM plays a vital role in production planning and production controlling function in the garment industry.

Incentive Systems – An Introduction

In the manufacturing industries, incentives are an important factor (financial or non-financial) that enables or/and motivates a course of action or counts as a reason for preferring one choice to the alternatives. Eventually, incentives' aim is providing value for money and contributing to organizational success.

The incentive scheme is considered as a driving force that produces higher productivity with the same resources available. Incentives can be classified according to the different ways in which they motivate agents to take a course of action. One common and useful taxonomy divides incentives into three broad classes:

1. Remunerative incentives or financial incentives are said to exist where a worker can expect some form of material reward, especially money, in exchange for acting in a way.

2. Moral incentives are said to exist where a choice is widely regarded as the right thing to do, or as particularly admirable, or where the failure to act in a certain way is condemned as indecent. A person acting on a moral incentive can expect a sense of self-esteem, and approval or even admiration from his community.

3. Coercive incentives are said to exist where a person can expect that the failure to act in a way will result in physical force being used against them (or their loved ones) by others.

Different Types of Incentive Systems:

Straight piece rate:

In the straight piece rate system, a worker is paid straight for the number of pieces which he produces per day. In this plan, quality may suffer. Usually, in the garment industry, this incentive system is in use, but this system promotes only productivity, not the quality which is a prime objective of garment manufacturing. This incentive system is much suited with the contract workers where management wishes to get maximum output with the limited number of working hours.

Straight piece rate with a guaranteed base wage:

A worker is paid straight for certain output set by management even if a worker produces less than the target level output. If a worker exceeds this target output, he is given wage in direct proportion to the number of pieces produced by him at the straight piece rate.

Differential piece rate system:
A system which suggests that there should be a separate rate for 70%, 100% and 120 % of target level. In this type of structure, fresher could hardly survive.

Halsey Plan:
According to the Halsey plan for incentive
$$W = R.T + (P/100)(S-T).R$$
Where,
W: wage of a worker
R: wage rate,
T: actual time taken to complete job,
P: percentage of profit shared with worker
S: std. time allowed.

Output standards are based upon previous production records available. Here management also shares a percentage of bonuses. Here the incentives are given based on time saved by the workers on a fix same wage rate. Here workers get motivates for doing the work with more efficiency but after a long time, workers use to be unsatisfied and demand more profit share through their works.

Rowan Plan:
According to the Rowan plan for incentive
$$W = R.T + ((S-T)/S).R.T$$
Unlike Halsey Plan gives a bonus on (S-T)/S, thus it can be employed even if the output standard is not very accurate. According to the Halsey plan, the workers will get more if they do more, but the quality of work might be distorted with the aim of more production. But with Rowan method, if the time saving is more than 50% of the standard allowed time then the bonus will be reduced. It means there are no benefits to do work with super high efficiency because it will affect the quality level of the production.

Bedaux Plan:
Like other plans, the minimum base wage is guaranteed. 'B' represents a unit of work. 1 B stands for 1 standard work minute and it

includes working time as well as time for rest. A worker earning "60 B" per hours reaches 100% of standard output or 100 % efficiency.

A bonus is paid to the worker who earns more than 60 B's in one hour. The bonus as in the original plan is 75% of the number of B's above 60 in one hour.

$$W = R.T + (Ns - Nt/60)(75/100).R$$

Emerson's Efficiency plan:

Workers with efficiency =67% to 100 %, an incentive is givens from 0 to 20%. For a 1% increase in output 1% increase in incentive.

Group Incentive Plan:

Equal distribution of cash or shares between the team or group employees involved in a work. When a production line hit the target output and produces more pieces than they will get paid incentive accordingly. In the group line supervisors, helpers, mechanics and even floor in-charge get the share of incentive amount. Share percentage is kept different (less) for supervisors and managers than operators and helpers.

Sources: Various web sources.

An Incentive Scheme for Sewing Operators

You know employees work in a factory for the monthly paycheck. If there is a chance for earning extra money through an extra effort and giving better performance, most of the workers will be happy in giving an extra effort. As a result, an employer gets more production in the same working hours. It will work only if you design your incentive scheme scientifically and if the incentive scheme is fair.

Here the term 'incentive' indicates the bonus amount or extra amount of money operators earn for their better performance.

Incentives scheme design based on the extra production

The name itself defines that bonus will be paid to sewing operators when operators produce more pieces than target production. Production target is set for the line and bonus is paid to the line. This is one types of group incentive plan where line output quantity will decide whether operators will get an incentive or not.

This is the most easy and smart way to design incentive plan for operators. Operators would be happy and motivated as they can count extra pieces produced by them at the end of the day. Hence, they can find out their incentive earnings easily. To keep it simple, incentive rate can be kept equal or less than the current average labor cost per garment. Reason, whatever extra garment is produced by the line will cost you nothing in term of labor payment, but you can save in overheads, electricity, overtime payments.

Managers, who feel operators and supervisors may cheat them in individual performance-based incentive scheme, can implement this scheme. Secondly factories, who do not track details of individual operator performance for their operators, can easily implement this incentive scheme without any difficulty.

Read following 5 steps to design production-based incentive scheme.

Step 1: Set production target (PT):

Industrial engineers need to set-up daily line's production target for operators to be eligible for earning incentives. Record daily line output from day one of loading the style. Once learning curve ends calculate average line output of following consecutive 3 days. For example, first 7 days line output is 40, 150, 250, 290, 300, 310, 305 pieces. You can expect daily average production of 300 pieces from the line without any bonus system. Now to get more production, announce bonus for extra production quantity. Operators will earn incentives when line make more than 300 pieces daily. Here incentives earning criteria is producing more than 300 units per day.

Target production can be defined more accurately by using line efficiency% history. If we use production quantity as the eligibility criteria, we need to capture production for each operation and each style. For the efficiency-based incentive calculation, you don't need to capture production data style-wise.

Here is the formula for calculating target quantity using line efficiency.

> Target production = (Available minutes in a line x Efficiency% / Garment SAM)

Step 2: Calculate daily average salary for the line (W):

It is considered that operators are paid a fixed salary. There are two ways to calculate daily average salary for the line.

1- Add up monthly salary of all operators working in the line. Then divide total amount by number of working days in a month to calculate average salary that is paid to the line daily.
2- Alternatively, you can first calculate daily wages of individual operators and sum up daily wages of those workers who work in the line to calculate line's daily wages.

Example: Lets' say 30 operators are working in a sewing line. 10 of them get ₹400.00, 10 of them get ₹500.00 and rest 10 operators get ₹600.00 daily. So, daily average wages for the whole line is ₹15000.00 (4000+5000+6000). To make the calculation easy multiply operator number by average operator salary. i.e., ₹500.00 x 30 = ₹15000.00

Step 3: Calculate Direct Labor cost per piece (C):

Divide daily total line wages (₹15000.00) by target production to calculate direct labor cost per piece. i.e., C=W/PT or C=15000.00/300 = ₹50.00

Step 4: Determine incentive rate per extra pieces produced (ET):

If the line performs as per target level, then line will produce average 300 piece daily. The cost of manufacturing per piece will be ₹50.00. Now, if line produces more pieces than your target you per piece direct labor cost will go down as because you are still paying ₹15000.00 to the line. You can set incentive rate two ways.

If line produces garments higher than 300 pieces, line will be eligible for earning bonus and the bonus rate is ₹50.00 for each extra garment. That is equal to the average labor cost per garment. In this case, labor cost per garment will remain same but your daily production will be increased, and order can be finished in less time.

Secondly, if you want to increase production and at the same time bring down your labor cost per piece, set bonus rate less than average labor cost per piece. You can set it as 80% of labor cost per garment i.e., ₹40.00 for each extra piece production.

Example: If the line produces 350 pieces per day, line will earn bonus for 50 pieces (350-300). Multiply extra pieces by incentive rate to calculate total bonus earned by the line.

If you use first method (i.e., ₹50.00 per garment) total bonus earning of the line in a day is equal to ₹2500.00. And if you use second method (i.e., ₹40.00 per garment) total bonus earning of the line in a day is equal to ₹2000.00.

Step 5: Distribution of incentive amount among operators:

You got total incentive earning for the line. You need to distribute to the operator in such a way that nobody gets demotivated.

Distribution of bonus can be designed in two ways –

- Distribute total amount equally to all operators. In this example, 30 operators earned bonus ₹2500.00. So, each operator will get bonus amount equal to ₹83.33.
- Second option: Instead of equal distribution, distribute bonus based on work content (SAM of the operation) ratio.

For example, if SAM of the garment is 15 minutes and SMV of two operations is 0.75 and 0.45 minutes and operations are done operator X and operator Y, respectively. As per SAM ratio -

Operator X will earn bonus = (2500 x 0.75)/15 = ₹ 125.00

operator Y will earn bonus = (2500 x 0.45)/15 = ₹ 75.00

This is one of the many methods of designing incentive scheme for the sewing operators. In this scheme, you can include other parameters, like minimum attendance in a week, quality performance of the line. When you are thinking of motivating operators through paying bonus amounts, you should also think of motivating other staff who are involved in line performance. Without their support, a sewing line cannot achieve the maximum performance. So, include line supervisors, helpers, and quality checker under the bonus scheme. Set bonus earning eligibility criteria for them as well. On reaching the target give them lump sum amount per day.

Incentive scheme implementation and Benefits

Are you still looking for the best scheme operator incentive? Do you know that many factories are taking benefit of the incentive system on every day?

I have recently visited a garment factory that has implemented preperformance-based incentive scheme for their sewing operators (salaried operators), line supervisors, quality checkers and floor in-charges. Within a span of six months company just doubled their monthly production with same manpower and other resources. In this section, I have discussed the following points -

- How the factory did it?
- What all benefits factory is experiencing by implementing incentive system?

Company profile:

The RMG factory is situated in the National Capital Region (NCR) of India. Under product profile, company makes ladies top with a lot of machine embroidery as well as hand embroidery work. Most of their sewing operators are migrants from other parts of the country. And most of them live far away from the factory location. The company provides a shuttle bus for the commuting to work. The company implemented incentive in a floor of 350 sewing machines.

The factory gained the following benefits after implementing performance-based incentive scheme

- Operator's efficiency level increased
- Many operators reached the target efficiency and earned lump sum amount weekly.

- Line efficiency improved more than 50%
- Monthly Production improved by two-fold in first six months of implementation
- Operators become more disciplined regarding spending quality time on producing pieces
- Operators agreed to do multiple jobs/ operations (which was a difficult task to convince an operator work on more than one operation throughout the day)
- Monthly capacity of the factory increased proportionate to the efficiency increase
- Considerable reduction in absenteeism percentage
- Above all, factory saves a lot of money after the successful implementation of incentive scheme.

How they did it?

The factory designed a simple incentive system to motivate their sewing operators. Under the incentive scheme, each operator can earn incentive irrespective of line's output once they meet following criteria

- An operator hits target efficiency and
- An operator produces more than the target.
- By reaching target efficiency operators earns a lump sum amount as jump bonus. Secondly, for each extra efficiency percentage after target efficiency, operator gets straight bonus.
- Operator incentive scheme is also linked with attendance of the operator. Like, if an operator works less than 5 days in a week will get no efficiency bonus.
- The incentive payment is made weekly basis.

To provide the correct amount of incentive, operator's individual production is captured, and daily operator wise efficiency is calculated by engineers. It was a very difficult task to measure the correct efficiency of everyone (operators) by manual data capturing.

This factory has installed a new technology, RFID based shop floor production monitoring and control system that helped engineer to measure real time performance of operators. This real-time production tracking system also provides a display device to each workstation on which operators can checks their day's production and total efficiency. Operator can even see their bonus amount earned on the display device.

Industrial Engineer's Digest

Attendance Bonus for Workers

This is a known fact that worker shortage is a big issue in the ready-made garment (RMG) industry in India. And a high percentage of operator absenteeism makes factory performance worse, in term of achieving factory's target efficiency and meeting deliveries on time.

To tackle an issue such as worker absenteeism, garment factories initiated giving attendance bonus to their employees. Even, factories compete with their neighbor factories to attract more workers with better salary and bonus amount and other benefits.

In the same zone where one factory is committing to pay bonus ₹400.00, others are giving bonus up to ₹650.00 per month for full attendance (year 2015).

Providing attendance bonus to workers is a good idea. Now question is - does your attendance bonus plan able to motivate your operators to be present full month?

It is true that operators come to the factory to earn money. If they get extra money just for coming full month, why they should be absent? It seems very easy to control absenteeism issue.

I discussed this issue with some factory IEs, who give attendance incentives to their workers (especially to the sewing operators). When it comes to the impact of the bonus system, I got different views from them. Read below to know what others think –

1. It makes a big difference.

Vineet Saini, the IE head of a large-scale garment manufacturing company in Gurugram, has experienced a positive impact of attendance bonus. Even in the day with bad weather, his lines were full of operators after implementing attendance bonus.

2. Not much difference.

Rajesh Kumar, IE manager from another garment export house, thinks operators who works in the garment factories are mostly migrated from other states of the country and they are living in industrial zone to earn money. Why should they do absent? Operators does not absent without any reasons. His problem is operator turnover. Instead of paying attendance bonus, management should take care of employees' issues for not coming to factory.

3. Attendance bonus is not effective.

Madhaiyan shared that in his factory they announced attendance bonus of ₹1000.00 per month for full attendance. And the same operator will be eligible to earn ₹1000.00 additional if he/she is present to work three months continuously. He said, even announcing such lucrative attendance bonus operators don't attend or fail to attend full month.

The above problem raised by Madhaiyan clears a point that money cannot motivate everyone. There are other factors also that HR needs to investigate to motivate operators and to reduce workers absenteeism rate.

Paul Collyer has mentioned a good point on the same forum. Why incentive bonus to be designed for 100% attendance only? "If at the start of the bonus period an employee is unavoidably absent, and the bonus lost then there is no incentive to attend for the remainder of the period."

Incremental attendance bonus will be more effective than 100% attendance bonus.

It is correct that when you are planning for giving employees an attendance bonus, it would be beneficial in getting increasing attendance rate. But the eligibility criteria for earning the attendance bonus should be designed such a way that would motivate your employees. Like giving an incremental bonus is a good idea.

Non-Productive Time (NPT)

What is Non-Productive Time (NPT)?

In the garment manufacturing, NPT is the short name of Non-Productive Time.

The time that is spent by an operator without producing any garment. For example, during a style changeover (line set-up time) operators are present in the line, but they do not stitch any garment.

In garment production, **non-productive time** is measured to understand how much standard time is lost by the factory during the shift hours. A report is made by recording machine downtime or lost time is called NPT Report. And this is an essential MIS report where factory value each standard minute of its operators.

Here, I will show you how garment factories capture NPT times and make the lost-time report.

In most of the garment factories, industrial engineers are responsible for recording and analyzing line-wise NPT for the factory. Normally, printed formats are used to record non-productive time. Factories normally record NPT time for the whole line as operator wise NPT data recording is a difficult task in manual tracking. In the manual tracking, only those non-productive time included where major time is lost, and which are visible to the data recorder.

NPT time is recorded to show management the reasons of low production in a day and lower line efficiency. A few examples of lost time categories are listed here.

- Line setting - time lost at the time of setting up the line for new styles. Once you have lost-time data, you can show it to your management as a reason for low productivity.
- Machine Breakdown - when machine related issues come during working hours.
- Cutting not available - No cutting is supplied to the production line. Operators are waiting for fresh loading
- Stitching Quality issue - When quality personnel found stitching defects (quality issue) in garments (in a bundle) stops making more garments till quality issues get resolved.
- Cutting quality problem – Cutting of parts do not meet the specification and operator spend time of fixing cutting issues.
- Power failure – No electricity to run machine

- Change of feeding plan – when a line stops working due to loading of a new style in the middle of currently running styles (order).

A sample NPT format has been shown below for your reference. Reasons of NPT may be different in your factory. So, make your own list of NPTs those occur frequently in your factory. For example, the operator sits idle due to delay in accessory receiving or trims approval may frequently occur in some factories and it may not happen at all in another factory.

			Non-Productive Time Tracking Format									
Line #	4							Date:	3rd-Jan-2020			
Line Supervisor:		Rajesh										
			Non-productive time Categories									
	Line Setting	Machine Breakdown	Cutting not available	Quality issue of stitching	Print/Emb. panel not received	Cutting quality problem	Power failure	Change of Feeding Plan	Others	Start Time	End time	No. of operators	Total Minutes	Supervisor Sign
1			√							9:30	10:10	12	480	Revised
2		√								12:10	12:30	1	20	Revised
3								√		15:30	15:40	20	200	Revised
4														
5														
6														
7														
8														
9														
10														
11														
12														
13														
14														
	Total												700	

Fig: Non-Productive Time Recording Format

How to capture NPT?

For capturing non-productive time of a production line, you can follow the below steps.

Provide a printed format to each line. Make one person responsible (line feeder, work study officer or line supervisor) to record lost time in total man minutes in the format.

During production hours whenever you see operators sit idle, find the reason of not having work (or not doing the task) and note down the start time and stop time of waiting, as shown in the above sample format.

In case multiple operators are sitting idle for the same reason multiply lost time by number of operators to calculate total man-minutes lost and record on your NPT format.

Lost time recorded under the category need to be approved by supervisor or authorized person.

At the end of the day calculate total lost time in each NPT category.

How to use NPT report for process improvement?

Once you have collected lost time report of a line for a period of a month, you can analyze and find which NPT parameters are occurring at higher frequency. You will get clear figure how many man-hours are lost due to presence of those NPT parameters in the process.

All these lost man-minutes can be converted into productive minutes by improving processes and strong production planning. Now you need to think upon reducing the major lost time parameters.

Do root cause analysis for occurring lost time and take an improvement project to reduce valuable standard hours.

Here, I have discussed 4 major non-productive time categories.

4 Major Non-Productive Time Categories

You may be aware that the average factory efficiency of the Indian garment factories lies between 30-45%. Why factories have that much low efficiency?

Low skilled operators in the factory are one of the reasons. But this is not the main reason for such low performance of the garment factories. Major reasons are non-productive activities occurred during production hours. Most of the garment manufacturing factories do not track the off-standard time and non-productive time (NPT). As a result, most of the factories are not aware how much efficiency are lost due to NPT.

I got a chance to study and analyze of non-productive times in six garment manufacturing factories situated in Northern part of India. All factories were medium size factory. Each factory track operator's non-productive time under 8-15 listed off standard categories. Factories tracked non-productive time using RFID based shop floor production tracking system, which ensures that most of the non-productive time had been captured.

Out of 15 non-productive categories, considering all six factories there are 4 topmost non-productive time categories in the shop floor that reduce line efficiency.

1. Operators waits for work,
2. Cutting not available for loading,
3. Alteration and repair work in production line and
4. Line setting

I call these non-productive time categories **'Efficiency Killers'**. Possible reasons of having high lost time in a factory, are explained here.

1. Waiting for work:

It means, in a line operators sit idle due to no feeding from the previous operators. Operator is ready to work but not getting enough bundles or pieces to keep her busy.

Possible reasons: Poor line balancing, operator absenteeism, quality issues. It is also observed that due to non-approval of trims makes operator to wait for a long time.

2. Cutting not available:

All operators may sit idle or a few operators at the start of the line sit idle for feeding of the next layer. Work is not available for the line.

Possible reasons: Insufficient fabric to cut and load, pending fabric approval to cut, delay in cutting, less cutting capacity or poor cutting plan compared to stitching floor capacity.

3. Alteration or Repair work:

When required stitch quality is not made at the first time, garment parts need to open and stitch it again. This task is called repair work or alteration. Sometimes damaged parts in a garment need to be replaced and re-stitched.

Possible reasons:
- An operator stitched a defective seam and garment has been given back to him for alteration.
- Shade variation in different component of the same garment and need to change parts to match shade,
- Operators are sitting idle due to no feeding, so they are given to do repair work for other style or alteration for same style,
- Quality check points are not available or insufficient quality checkers in a line.

4. Line setting:

During setting up a line for a new style, many operator wait for the first bundle. Machine arrangement for the new style, checking machine set-up, and arranging the workstation takes time.

Possible Reasons:
- Frequent change of styles due to small order run increase efficiency losses. It is observed that 4-7 styles are loaded in a month to a line.
- Due to shipment pressure or poor planning, sometimes line supervisors need to stop a line without finishing current style and a new style is loaded without prior planning and resources. These cause major loss of standard time.
- When a factory runs single style to all lines on the floor and complete stitching in all lines on the same day. In this situation line setting is happened in multiple lines on the same day which needs extra line-setting time than normal line setting.
- Inadequate fabric for the style – lines need to be stopped, and other style is loaded until fabric is sourced and cut.

This list of major NPT types and root causes for having high NPT hours is prepared based on my observation and study about major non-productive time categories. In your factory, the critical non-productive time categories might be different. Through measuring it and data capturing, you will know the fact of NPT categories.

Why Measure Non-Productive Time of a Line?

In the previous article, I have explained what the non-productive time is and how to measure non-productive time. I have also discussed about the major causes of lost time.

In this article, I will discuss the reasons why you must measure and find actual non-productive time in the production floor. You should do so to achieve the followings -

- To know the fact of the shop floor that how productive hours are lost and how many hours are lost on every day. Management just looks at the daily line efficiency data and labor productivity. But why efficiency and productivity are low in your factory? Major reasons are lost time.
- You can only manage (reduce) lost time in a line when you measure it, and you have enough data for decision making. That is why you need to measure non-productive time.
- Unmeasured non-productive time is the potential area for productivity improvement. Productivity can be improved by reducing lost time.
- By tracking lost time and reducing it, you can reduce the production lead time. At least you can be in position in shipping goods on time.

I would like to share one case study. One trouser manufacturing company had undertaken a project to reduce lost time in their production lines. After two months of the project initiation, by reducing lost-time in different lost-time categories, the company has improved line productivity more than 10%.

You may be thinking how this is possible? Let me assure you that this improvement is possible provided you know how much total time is lost in your factory and you know how to convert those lost time into productive time. Track improvement opportunities within your reach. Let me share how they did it.

Earlier they do not have any measure for lost time data. They implemented real-time production system and instructed their operators to record lost time when they do not get work and when they sit idle for any other reasons. Operators are also instructed to enter lost-time reasons in the real-time production tracking device. They have extensively captured lost time data in different categories (as occurred). They analyzed data for one month. They got clear picture of the major lost times categories and number of hours factory is wasting in each of the categories. They prepared a plan to reduce top level loss times categories and found good result in one month.

In the garment industry, many garment factories do not know how much productive time they are losing every day. In case sewing operators and helpers are not working full 8 hours that means you have loss time. Being present in the floor for 8 hours and working for 8 hours are two different things.

Utilize operator's time as much as you can. There is no better alternative than just stopping operators sitting idle, to improve operator productivity. Non-productive time such as waiting for work, machine breakdown, power failure and repair work, kill your productivity. Start eliminating such non-productive time as much as possible. To start work on this point, you need to track non-productive time data in different categories. Once you have the analysis and Pareto of non-productive time you can think and plan on reducing it.

Do you measure non-productive time of the sewing lines in your factory? If you do, you are on the right track in improving resources utilization and improving productivity. If not, you must start measuring non-productive time on the shop floor.

What is Needle Downtime?

The term 'needle downtime (NDT)' is often used in the apparel industry to express total time or percentage of total work time of a machine runs in sewing garment. This is needle running time.

The word 'downtime' (or 'downtime') sometimes makes confusion, and many may mean it as machine idle time and is considered as one of kind of non-productive time.

When I tried to define the term needle downtime, I also had confusion about the correct meaning of this phrase. For the clarification, I seek help from industry experts and asked for their opinions.

Dr. Prabir Jana has explained this term very well. "Traditionally 'downtime' word is associated with 'anything not working'. However, in case of sewing machine, it is just opposite 'sewing machine works when a needle is down'. He suggested use the word 'needle running time' to avoid confusion."

Mr. Keerthi Abe explained Needle downtime as 'Needle up & down going time'.

Needle downtime is considered as one of the performance measures (KPI) for sewing floor as well as sewing factory performance.

I hope you understand the meaning of needle downtime in the garment manufacturing.

The higher the needle downtime, better is the machine utilization. By improving needle downtime, factories can improve factory's performance in terms of line efficiency and labor productivity. As per some published study, needle downtime in the garment manufacturing industry is less than 15% of the total worked time.

Industrial Engineer's Digest

How to Calculate Needle Downtime?

In the previous article, the definition of ***needle downtime*** is explained. In this article, I will show you how to calculate needle downtime. Needle downtime is also referred to as needle time. Needle time is the total time a machine runs during the day. We can define it as the total time spent on stitching garments.

If you observe an operator while she is stitching a garment, you will see that the operator does not run the machine continuously. She picks up garments (or garment parts), places it on the table, aligns pieces, put the pieces under pressure foot, stitches few centimeters of the seam, re-aligns plies, and stitches the piece. In one operation cycle time - there is material handling time and stitching time (needle downtime).

If you observe sewing operators' activities throughout the day, you will find they

- stitch garments (needle running time)
- sit idle when they have no work to do, or for other reasons of lost time
- speak with co-workers
- move from one workstation an another
- take break
- go to the restroom
- they chat with quality control person and line supervisor
- and many more uncertain things may happen

From these many activities, you need to measure only needle downtime – i.e., when machine needle moves up and down. You can use one of these two ways to measure needle downtime.

- Statistical data analysis (Work sampling method)
- Operation cycle time analysis

Work sampling method

It is not possible to observe and record needle time for all operations throughout the day. By work sampling method you can find an occurrence of different activities in terms of percentage. Using work sampling method, find the percentage of machine running time out of total hours worked. From this percentage data, you can calculate how much time is spent on running machine by operators. Remember, this is statistical data analysis and not actual needle time. For actual needle downtime follow the second method.

Measure needle downtime by cycle time analysis

You might know how to do cycle time analysis. While you perform time study for an operation, you capture time for different elements of the operation cycle. Calculate machine running time (stitching time) out of these elements. This is the needle time per operation (for stitching operation of one piece).

Let first find needle time for an individual workstation.

- Select one workstation and an operator.
- Write list of operations she is doing. Do time study for single operation first
- Do time study for all elements of the operation cycle
- Find time for machine running element (stitching) from the time study sheet
- Study at least 25 operation cycles for same operation
- Find average time for stitching element of 25 studies
- Count total pieces made by her (operation wise quantity if she is doing more than one operation)

> Total needle downtime in a day = (No. of pieces made x Stitching time per piece)

Let us say, one operator is doing collar run stitch operation of a shirt. The calculated average machine running time (needle downtime) is 12 seconds. She has produced 400 collars (only collar run stitch). Therefore, calculated needle downtime would be 4800 seconds (or 80 minutes) in a day. [400 pieces x 12 seconds=4800 seconds]

Calculate needle downtime in the percentage of total hours worked by dividing 80 minutes by 8 hours. It is (80x100/480) % = 18.33%

In this example, we are not considering whether this operator worked only one operations or multiple operations and if she did repair work in the same workstation. To know actual needle downtime, you must include all activities an operator did by stitching machine.

143

The Concept of Idle Time

You might have heard about the *idle time* associated in apparel production. Idle time is considered as one kind of loss time (non-productive time).

Garment factories those follow progressive bundle production system must have experienced a situation when operators wait for the work after completing one bundle or one piece. This waiting time is *Idle Time*. The situation when operator is ready for doing work, machine is in working condition, she is just waiting for work to feed to her, is called as idle time.

This idle time or non-working time can be significant such that the operator's true efficiency is reduced in proportion to the idle time between bundles. The cumulative of total idle time can be huge and factories may lose many productive hours every day.

For an example, in a 40 machines line 10 operators are working without additional bundle (WIP) but the bundle they are working currently. Assume that all of them wait for 30 seconds for the next bundle (or for the next garment in single piece flow movement). If they produce 400 pieces in a day (40 bundles), total ideal time for each operator will be 30*40 = 1200 seconds = 20 minutes, which is equal to 4% of shift time.

Why it is essential to record and analysis idle time in your factory?

Factories strive to improve their performance (efficiency) but sometimes they are lost in finding a way for efficiency improvement. For them working on reducing *idle time* is an opportunity for improving factory efficiency.

Most of the garment factories ignore measuring the idle time. They just let it go. One reason might be management wants to hide their inefficiencies in not to have a balanced line with enough WIP at each workstation. You can only reduce idle time when you have data of total idle time.

I have seen one factory that captures operators idle time and measures operators' on-standard efficiency. They capture actual bundle work time in a day. Later they subtract the total bundle work time from the shift work time to calculate operators' idle time.

Therbligs and 18 Motions Name with Symbols

What is Therblig?

Therbligs refer primarily to the motions of the human body at the workplace and to the mental activities associated with them. Hand and Eye motions. The human body movements are divided into divisions of movements or group of movements for the micromotion study. These divisions of movements or groups of movements are known as Therbligs.

Micromotion study is done for operation with a very short cycle time that repeated many times.

The origin of the Therblig term is very interesting. **Therblig** is an anagram of **Gilbreth** (Frank B. Gilbreth), the founder of the motion study. The divisions of the human body movements (activity) were devised by Frank B. Gilbreth.

18 Motions of Therbligs, Symbols, Names, Abbreviation

Each therblig has a specific color, symbols, and letter for recording purpose. The Therblig symbols, the name of the motion and abbreviation (letter) are shown in the below chart.

Symbol	Name	Abbreviation
⊂⊃	Search	Sh
⊂⊙⊃	Find	F
→	Select	ST
∩	Grasp	G
⊓	Hold	H
	Transport Loaded	TL
∪	Transport Empty	TE
9	Position	P

Symbol	Name	Abbr.
#	Assemble	A
U	Use	U
⌗	Disassemble	DA
◯	Inspect	I
8	Preposition	PP
⌒	Release Load	RL
⌒ₒ	Unavoidable Delay	UD
	Avoidable Delay	AD
⌐	Plan	Pn
⌐	Rest	R

Resources:

1. Introduction to Work Study, 3rd edition
2. https://en.wikipedia.org/wiki/Therblig

What is TMU in Time and Motion Study?

TMU stands for **Time Measuring Unit**.

In Methods Time measurement (MTM) system, a predetermined time is given to each motion. In motion study, the time for each basic element is given in units of TMU. In the following table, example of TMU value for a motion is shown against different distances.

Table: TMU value for motion study (Source: SPD book)

SPD Table	Ref Code	Description	Distance in Inches	TMU
S01	S101	P/u single part, take to lap or to table (4,8,12,18,24,30)	4	24
S01	S101		8	32
S01	S101		12	37
S01	S101		18	45
S01	S101		24	53
S01	S101		30	61

How many TMU is equal to one seconds and one minute? Conversion of TMU to hours, minute and second

1 TMU = 0.000010 hour = 0.00060 Minute = 0.036 Second
Or conversely
1 second = 27.78 TMU
1 minute = 1667 TMU
1 hour = 100,000 TMU

Application of TMU in the apparel industry

For establishing the **standard time (SAM)** of sewing operations by using the PMTS system, motion codes are applied for each motion (movements) and each motion carries time in TMU. In Standard time calculation, initially, standard time is measured and calculated in TMU. Later, the total TMU is converted into minutes.

What is a Learning Curve?

In the apparel manufacturing, a learning curve is made for the sewing operators' performance build up day by day after they start working in a new style.

A learning curve of an operator is the performance growth of getting experience (time per unit) in a task. The learning curve can be drawn time per unit against cumulative repetitions or efficiency at work against cumulative repetitions.

When a new style is loaded to a line, sewing operators (machinists) take time to reach their maximum speed (or normal speed). A sewing operator takes higher time to stitch a garment when she starts working on a new style. After doing the same task for many times, the time taken to stitch the garment, or an operation gets reduced. It may be due to new operation (job), a different machine with different settings and new guide or a different kind material need to handle.

For example, the optimum output of a line is 500 pieces/day. 1st day it may produce only 100 pieces, 2nd day 250, 3rd day 450 and on 4th day onward it produces 500 garments. Same things applied to individual operators.

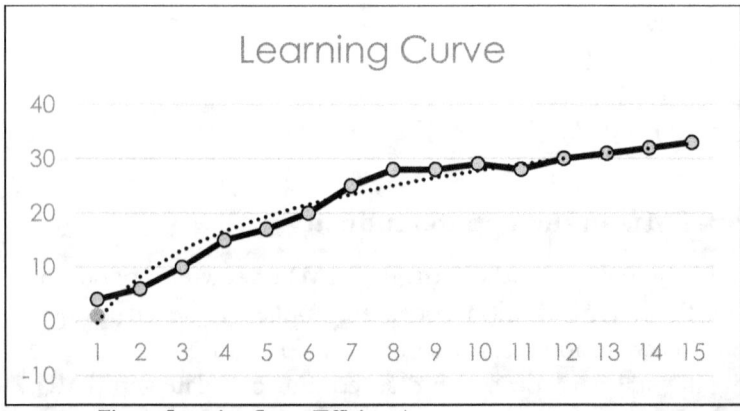

Figure: Learning Curve (Efficiency)

Individual operator's learning curve is normally prepared at the operator training stage. But you can even prepare learning curve for each sewing operator working in the production line. I have seen some factories prepare all operators learning curve.

Normally, the learning curve is prepared for the production line. This is easier to develop line wise learning curve when you do not have individual employee efficiency data. See the learning curve of a typical production and typical garment style.

In case you have short and long run orders, develop different learning curves for different order volumes. The reason, for short-run orders, you may not reach to line's pick efficiency level within the time production is over.

Use of Learning Curve:

Following are few reasons, a learning curve is made by industrial engineers.

- Industrial engineers use learning curve data to calculate daily production target in the initial production days of the new style.
- In production planning and target setting, the learning curve is used to plan day wise production target.
- A learning curve is also used to derive lead time of an order.
- For the incentive calculation in the learning period, the learning curve is used. If you have incentive scheme and you want to give incentive to your workers based on their efficiency level, learning curve helps in designing the target efficiency (eligibility level to earn incentive) in the initial days of production loading.
- In operator training stage, learning curve is prepared to assess employees learning growth.

You can develop learning curves for each production line in your factory and develop a separate learning curve for different products and different order volume. Learning curve is also presented in term of line efficiency% (learning curve efficiency).

Why Develop Database for Learning Curve?

I was asked this question, "Is there any benefit of developing a learning curve database in such case where style does not repeat?"

This is a good question. And you should also know why one should develop a database for learning curve, when there is no chance of getting repeat orders. In the following paragraphs, I have briefly explained the benefits of creating a database for learning curve for lines. Though I have mentioned creating a database for the learning curve, this thing is true for preparing the database for other areas too.

Based on the Pareto's principle (80-20 rule), you will find 80% similar operations in new styles, though you are not getting repeat order of a style (I am assuming that you will be working same product or similar product range). In production floor, the same machines will be used to make the new styles - 80% machine will be same as used in earlier styles. Also, most of your sewing operators are the same. So, in the most cases in a new style you will not be doing something completely different operations.

Create a database for daily efficiency figures (line efficiency as well as individual operator efficiency) for few styles during learning curve (initial days of production start of a style). If you review the data, you will find similar progress (trends) of efficiency after loading in most of the styles.

The purpose of making this database is to use for planning production target on the initial production days. And secondly planning incentive scheme for a worker on those days. You have no other way but referring past data to plan things in the initial days.

Later at any point of time, you can check the database for analyzing the performance of the line. The database can be used to compare with running orders. The comparison can be done in between sewing lines. By having the study of a longer period (suppose 6 months to one year) you can set your benchmark for the learning curve.

To get more precise data, you can make a separate database for different apparel products. This data can be used for planning future orders.

Here developing database means recording day wise operator efficiency and line efficiency for the complete style run and then making a separate report for initial production days up to learning curve. You can also make trend chart for every style.

Part-II

Data Capturing, Calculations and Reports

Production Reports Made by IE Department

Reports are categorized as daily, weekly, and monthly reports. All these reports are important to track the factory's performance from micro to macro levels.

You might be using some of or all the following reports. If you are not using these report and you are planning to set up IE department in your factory or if you have just established an IE department, you can implement these reports.

Daily Reports

Reports those are made on daily basis are known as daily reports. IE department prepares following reports on every day.

- Daily production report
- Line, factory, and group efficiency report
- Line, factory, and group Lost time report
- Factory wise running style Costing report
- Daily factory KPI report
- Line, factory, and group WIP report
- Individual sewing floor efficiency/Performance report includes operator and helpers
- Individual cutting and finishing efficiency/Performance report includes helpers and checker

Weekly Reports

Weekly reports made by IE department

- Consolidated weekly production against planned
- Consolidated weekly efficiency against planned
- Consolidated weekly Line WIP report
- Consolidated incentive amount working report

Monthly Reports

IE department prepares following reports once in a month.

- Efficiency reconciliation report

- Lost time reconciliation report
- Completed styles reconciliation report
- Individual efficiency report for operator/helper/checker
- Machinery requirement report
- Training center progress report
- Monthly MIS meet report – Factory KPIs analysis
- Thread consumption report by style
- Folder and attachment requirement report for new styles
- Line wise operators' skill matrix report – once in 3 months
- Group incentive report

Name of the reports may be sounding differently to many. Name of the report is not important but the measures that you need to capture and prepare to understand where your factory stands.

These production reports can be by garment factories even if they do not have industrial engineering department.

Necessary Production Reports for a Garment Factory

In a garment factory, a production manager's day starts with checking the production reports and WIP level in the floor. Production manager enters to the factory floor with these questions -

- How many garments are made on the last production day by each line?
- Compare actual production with production target given to the floor in-charge and line supervisors.
- What were the issues for low productions?
- Why too much quality issues?
- Who is responsible for not achieving yesterday's target?

Production managers get answers to their question in reports with data.

In this industry, line supervisors are very smart to tell you numbers to show good line performance and hide their shortfall. So, instead of discussion and verbal reports, most of the managers believe in written and printed reports.

To know the factory's day to day performance very well what all production and production-related reports must be made and reported to higher management? Each factory has its own set of policies and system to run their factory. So, list of production reports may vary company to company.

There are few common reports; those are widely used in the garment manufacturing and these reports are essential to know your business well. Here are some of those reports related to production. I have explained the purpose of making those reports. Few reports are information-based, and others are analysis and performance-based.

1. Operator Attendance Report:

This report explains how many sewing operators; helpers are present today in each line. From this report managers quickly assess, what would be today's production target. Which lines have a shortage of operators and where there are enough operators, and from where some operators can be shifted to another line to fill all lines and run production smoothly. Operator attendance report is an important one to floor in-

2. Daily Production Report (DPR):

charge. Yesterday's operator's clock-in and clock-out report is also important to see whether all operators worked full time or not.

Daily Production Report contains yesterday's production (last day's production) records line wise and style-wise. Production information in term of –

- Total cuttings are loaded to each line yesterday and total loading done to a from loading of the style.
- Total pieces were stitched yesterday and the cumulative production till-date of all styles those are currently loaded.
- Total pieces have been dispatched to finishing or washing department.
- From daily production report, managers can assess, whether a line is producing as per target or production is getting delayed.

3. Hourly Production Report:

This report carries information of today's hourly production data. In this report, line's output is being updated hourly or bi-hourly. This report helps line supervisors to track operators when line output goes down. Where production data is recorded manually, only line output data is captured and displayed on the production board. But factories those use real-time production tracking system, operator wise and operation wise hourly production data can be viewed. I have shown one such report for example in the following table. Production counting is done based on number of bundle completed and number of pieces in each bundle in the hourly slots.

Table: Hourly production report table (Operator and Operation wise)

	Operation	Operator	Hourly production (H=Hour)									Total Qty
			H1	H2	H3	H4	H5	H6	H7	H8	H9	
1	Placket press	Lalit	10	25	50	20	30	20	25	50		240
2	Gathering lower	Alok	25	50	45	45	35	40	50	45		335
3	Dart making	Pushpa	30	40	40	50	35	35	40	40		290
4	Yoke attach	Ankur	30	50	40	40	35	35	50	40		300
5	Yoke O/L	Sandeep	20	50	40	35	25	35	50	40		285
6	Yoke T/S	Sagar	25	45	45	15	15	45	45	45		290
7	Placket run stitch	Pooja	5	20	20	25	20	20	20	20		140

4. Efficiency and Productivity Report:

The daily production report provides actual output numbers of each line and each style. The line output is directly proportional to number of manpower in a line. If you have two different styles (of different SAM) running in two lines, the performance of that line cannot be compared with output quantity. The easy way to check the line's performance is measuring line efficiency and machine productivity irrespective of styles, total number of manpower and number of machine used.

5. Manpower and Machine utilization Report:

Manpower and machines are the primary resources of a manufacturing unit. A manager needs to check resources to check how company resources are get utilized. The factory should not have excess manpower on the floor. On the other hand, the factory must have minimum required manpower to a factory to produce goods according to the plan. Both the manpower and machines are the cost to the company. It is important to check this report daily basis.

This report may include employees who were present in overtime work hours.

6. Garment Inspection Report:

Inspection reports come under quality report. But garment inspection report is also one important report for the production manager and top management to monitor the stitching quality of products. As all produced garments are not acceptable if a certain quality standards are not maintained. It is the production department's responsibility to produce quality products. Inspection report displays line wise defects per hundred units (DHU) and percentage defective data, total pieces send for alteration and repair.

7. Repair and Rejection Report:

This report contains information such as – style wise and color wise defective garments produced by lines. Defect-wise the number of garments is sent to repair section and number of the garments received from repair section after the alteration. This information is essential because in case you have a few damaged or rejected garments, you can cut and sew fresh garments and replace those damaged garments prior to final inspection.

8. Cutting Production Report:

Cutting department's production status is also an important report to be investigated by the production manager. Style wise and color wise cut quantity is recoded in this report. Sewing line can be fed with cuttings only when there are enough cuttings in the cutting WIP. Looking into cutting plan what all fabrics are going to cut today for the following day's feeding.

9. Material Inventory Report:

Once the status of production, efficiency, and quality report part is over, managers check material availability status. You know that unavailability of materials is one of the major reasons for shipment delay and breaking the production schedule. This report carries information such as fabric and trims in-house status in details with expected in house date for balance materials. Managers chase with merchandising and sourcing department for sourcing material on time.

I have shared list of production reports. Now it is your turn to implement these basic reports in your factory and standardize these reports.

Tips to Make Daily Production Report (DPR) Quickly?

Is your morning task to make Daily Production Report (DPR)? Does it take too much time of yours?

Why it takes too much time to finish Daily Production Report? Did you try to analysis what stop you making the report on time?

I am sharing my personal experience on this. In the recent past, I have implemented MIS system (data capturing formats and report generations) in a garment factory. Data operator was allocated to make daily production report in the morning as early as possible. During the implementation of report making, we found many obstructions that cause the delay in report making. Scheduled time for report distribution was at 9:30 am (Factory starts at 9:00 am).

Followings are few obstructions I had observed -

- Computer unavailability,
- Late receiving production data from different departments,
- Department head reached office lately,
- Morning meeting keep HODs busy,
- Unskilled data operator,
- Data operator does too much experiment on the excel sheet while making report etc.

In many factories, DPR is made by industrial engineering department. Either the report is made by IE or by data operator. Whoever is making the report, following tips would help them to be on track.

Tips to reduce report making time

1. Allocate a computer to data operator in the morning

In the small size factory, a full-time computer is not given to the IE or data operator. If you are working on a shared computer, make sure that you get the computer in morning for half an hour.

2. Use simple formats for data collection

To prepare the DPR, the second thing you need after the computer is production information from all the production departments.

Such as cutting, sewing, packing, and finishing departments. Give them a simple printed format to make department wise production report for you.

3. Use template for feeding data in Excel sheet

Use formatted spreadsheet. Do not try analysis data daily. Instead, prepare templates for charts and graphs that you need for reporting. In the morning just collect and enter data in the specific tables.

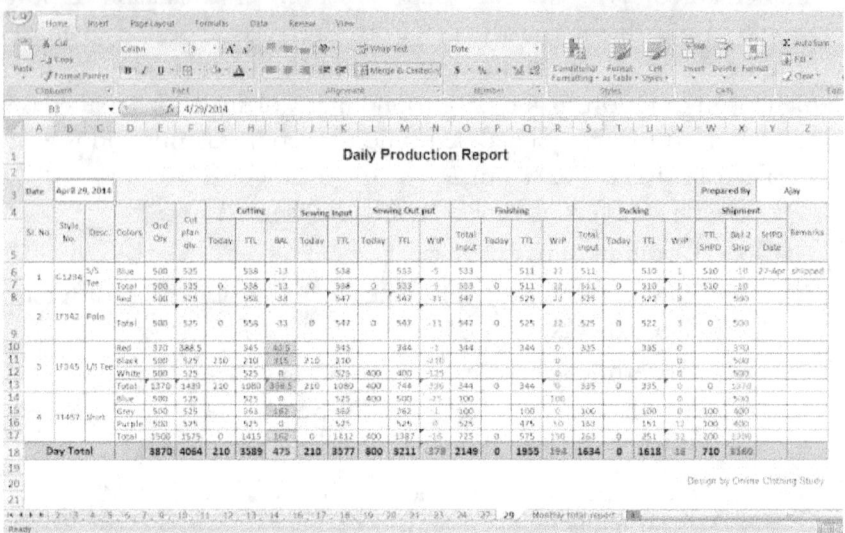

Figure: A sample daily production report format

4. One task at a time

Keep other tasks aside in the morning and concentrate on the data entry.

5. File and Folder Management

Work on a minimum number of files for daily reports. Single file is the best for a month. Create date-wise separate worksheet in the same file. Save all production report files into the same folder. Create a shortcut of the folder in desktop. You can also share and save report folder in a shared location. This way you can work on the production report file from other computers (provided that other computers are in local network).

6. Master yourself in Excel

In case you are slow in working on Excel table, do not know much about shortcut keys, formatting of tables, cross-checking of calculation, the basic formula for auto calculation, practice when you get time. Ask for help and learn from others. You can learn Excel by watching YouTube videos.

By following the above tips, you can surely reduce report making time. Download the sample daily production report with instructions and Monthly Report sheet by clicking this link https://goo.gl/aepT1C.

If you are reading printed version, use the following URL for getting the list of downloadable files and from the list download Daily Production Report tempalte
https://www.onlineclothingstudy.com/2015/10/list-of-download-able-filestemplates-at.html

Hourly Production Report

In a garment factory, on the shop floor, line supervisors get production target for the day. Production target means the number of garments needs to be produced by the end of the shift using given resources (man, machine, and hours).

The daily production target calculation part is easy but producing the target quantity from a line is not. So, you need an effective tool to control daily production at every hour. And the basic reporting tool used by garment factories is Hourly production report. Also called as Production by Hour Report. Here, I have explained two things

- What is an hourly production report?
- How to use hourly production report to control daily production?

What is an hourly production report?

A report in which hourly production quantity of a line (line output) or/and hourly production of all operations and operators are captured is known as **an hourly production report**. For the current day, hourly production is displayed on a whiteboard.

After the first hour, it shows production on that hour and cumulative production till the hour (production so far). The hourly report format also mentions daily production target, manpower, product SAM etc. The hourly target can be also mentioned on the whiteboard.

Hourly production report for more than one line on a floor can be shown in a single display board or individual hourly production report display can be made for all lines.

Traditionally, one helper (or work study person) goes to the end-of-line checker or output operator in each line and collect the garment quantity they made on that hour. In some factories, operators used to write down produced pieces (bundle quantity) on a piece of paper. Data operator collects production data from that piece of paper and writes the quantity in a production board. In case operators do not write production on a paper, data operator manually counts stitched pieces.

Data collection for hourly report

Depending on data requirement, you can make hourly reports one of the following ways.

Industrial Engineer's Digest

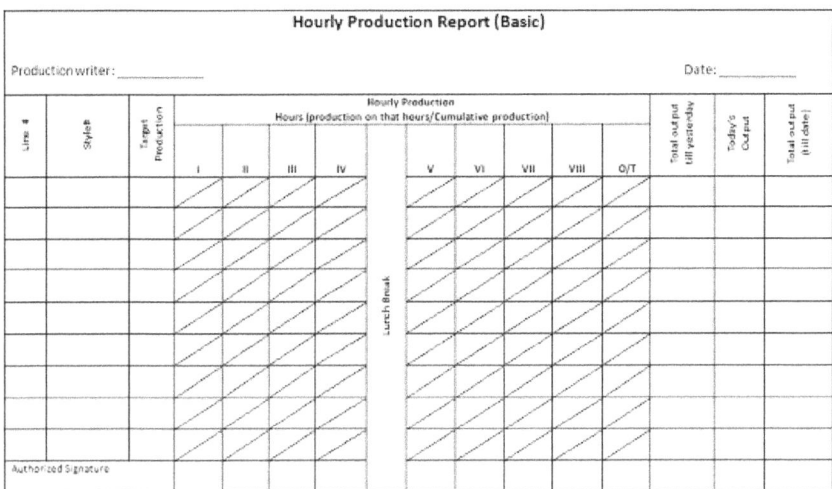

Fig-1: Hourly production report format

1. Hourly report for line output:

Only line output is captured and displayed. By making this hourly report, you will get hourly production quantity. But it would not be possible to know how much pieces are made by individual operators inside the line in other operations.

Table-1: Production by hour report (For all operations)

Hourly Production Report												
Date: _____ Line# _____												
			Hourly production (H=Hour)									
Emp#	Operator	Operation	H1	H2	H3	H4	H5	H6	H7	H8	H9	Total Qty
11002	Lata	W/B atta to lining	40	81	115	108	44					388
21220	Kabita	W/B top stitch	40	106	99	108	88					441
21221	Sikha	WB Fit lbl attach	0	93	93	77	22					285
22312	Anil	WB WC lbl attach	40	44	36	0	0					120
11231	Sumit	Parts O/L	80	167	55	100	192					594
11321	Susmita	Fly Iron	0	0	128	0	0					128
12341	Anchal	Fly ready right	62	73	40	73	0					248

2. Hourly report for all operators working on a line:

This hourly report gives production data for all operations (and all operators). This one is an in-depth report for data analysis to see line balancing, finding low performers in the line. See the Table-1, for this kind of report.

Control daily production using hourly production report

The fact:
Factories normally work for 8 hours a day. For planning and estimating line capacity (production target), equal production is considered in every hour. Though you will get less output in morning hours compared to average hourly production.

Line supervisors use to check line output hourly, but they do not question to their operators why production is less/not matching the hourly target quantity. They hope that by the evening they can meet target quantity. But it is found that at the end of shift they are not near to the target quantity. So, to meet the daily target they plan for overtime working (OT).

The Way:
Circumstance may change in every minute. So, your line output is. How you can control your production and achieve the target by end of the day. If you get hourly output from the line output and from the critical operations (bottleneck operations), you can assess how much pieces can be expected in eight hours. To speed up production you can plan for the balanced hours to meet the daily target.

In case you do not know how many pieces are made in first two hours or by half day, you could not assess how many pieces can be made in 8 hours.

I would suggest you capture hourly production data (for all operations or at least for critical operations in the line) and analyze the production report. By checking actual production every hour, you can estimate expected production in 8 hours. This report will help you to identify line WIP, and bottleneck operations. Even you can check line balancing in every hour.

Bi-hourly production report

Instead of capturing production hourly, when the production quantity is captured in every two hours, the report is called as bi-hourly report. The method of making hourly production report explained above can be used for making Bi-hourly production report. If you are not yet capturing hourly data, you can start with making bi-hourly production report.

Calculating Labor Cost Using Work Measurement

Question: *Can you please explain how we would use work measurement data in garment costing?*

Time study data is known as work measurement data. From time study we calculate the **standard time** (SAM) of various activities. In garment production, jobs are called operations.

Using work measurement data, you can estimate only direct labor cost of garment making. Other cost parts of the total garment cost such as factory overheads and material costs need to be calculated separately. Here, I will show you how to use work measurement data to find labor cost.

Be noted the cutting cost and finishing labor cost have not been included here. These costs are considered as factory overheads. I am not saying that you cannot measure the work content of cutting and finishing operations. You can also calculate direct labor cost for these sections.

Procedure to calculate labor cost

Step-1. Measure garment SAM:

First, calculate garment sewing wok content in standard minutes. You can use the Time Study method or synthetic data for calculating standard minute of the garment sample.

In case you need to learn the method of calculating garment SAM, read the article *"How to calculate SAM of a Garment Operation"*. So, you know how to get garment sewing SAM.

To calculate labor cost, you need additional information such as
- Daily labor wages (You can take the actual data)
- Line efficiency

Step-2. Calculate Daily labor wages:

Calculate the minimum labor wages of your region or check with your human resource (HR) department for daily wages of skilled labors (sewing operators). From the daily wages, calculate per minute labor cost. Considering shift time is 8 hours, which is equal to 480 minutes.

Let us say daily labor wages is USD 6.0 and
Calculated per minute labor cost is 6.0/480 = USD 0.0125

Step-3. Calculate Line Efficiency:

Calculate the line efficiency of your sewing lines. Line efficiency data is also calculated based on the Garment SAM. In another article, I have explained line efficiency calculation method. You need total produced minutes by a line and total hours worked by the line.

If you know the efficiency of your sewing lines, use that data. Let's say one line is running at 50% efficiency. We will use this data for labor cost calculation. When a line works at higher efficiency, it will produce more garment compared to a line working at lower efficiency level. It means, when a line efficiency increases, labor cost per garment will reduce.

Step-4. Calculate direct labor cost

Use the following formula to calculate direct labor cost per garment. In the formula we used three parameters to derive garment cost – garment SAM, line efficiency and labor wages.

> Labor cost per garment = (Garment SAM x Labor cost per minute)/ Line Efficiency%

Example: Let's say the garment SAM is 20 minutes. Line efficiency 50% and Labor wages of skilled labors are $0.125 per minute. Therefore, direct labor cost of the garment

= US$ (20 x 0.0125)/50%
= $0.50

All the above data can be found easily by the industrial engineers. Follow the above guides and calculate labor cost in garment manufacturing. If you are new in this field, learn the SAM calculation method and learn the method of calculating line efficiency.

Measuring Productivity in a Garment Factory

What is Productivity?

Manufacturing Productivity can be defined as "Output" compared to "Input".

According to Marsh, Bruce (2002), productivity is a measure of the efficiency and effectiveness to which organizational resources (inputs) are utilized for the creation of products and/or services (outputs). Productivity measurement is both a measure of input utilization and an assessment as to whether input utilization is growing faster than output[1].

In the case of a garment manufacturing factory, "output" can be taken as total number of products manufactured, where "input" is the people, machinery and factory resources required to produce those products in a day (within the given time frame). The key to cost-effective improvements is output and productivity. This is to ensure that the relationship between input and output is properly balanced.

There is little to be gained from an increase in output if it comes only as a result of a major increase in input. Indeed, in an ideal situation, "input" should be controlled and minimized whilst "output" is maximized.

Productivity calculation

The productivity formula can be written as the following -

> Labor productivity= (Total line output in pieces / Total labor input in manpower)

Let's calculate the labor productivity of a sewing production line. Let's assume,
1. Total garment production in a day = 800 pieces
2. Total number of labors (operators + helpers) = 40
3. Shift hours = 8 hours (working hours in a day)

Therefore, calculated labor productivity per 8 hours shift
= (800/40) Pieces =20 garments per labor per shift.

---

[1] A Consolidated Approach to Productivity Assessment, Journal of Industrial Technology

Benefits of measuring productivity

Higher productivity gives more output from the line from the same number of workers, in the same time frame. This, in turn, improves "overhead recovery" related to factory costs, such as electricity and fuel. Overheads are fixed costs within that time frame. So, the more garments are produced in the given time frame, the less overhead allocation per product, which, in turn, reduces the cost of each individual item and therefore improves competitive edge.

Dr. Bheda, in his book *"Managing Productivity in the Apparel Industry"* explained the different ways of measuring productivity. Productivity can be expressed in many ways but mostly productivity is measured as labor productivity, machine productivity or value productivity. These three ways of measuring the shop floor productivity used in a garment unit.

- Labor productivity - Output per labor (direct +indirect) in the given time frame (in pieces)
- Machine productivity - Output per machine in the given time frame (in pieces)
- Value productivity - Total value of output in the given time frame.

Labor productivity and machine productivity calculation methods will be explained in the following two articles.

How to Measure Labor Productivity?

In the earlier section, I have explained the definition of productivity. Productivity is the ratio of output and input. In this article, I will show you how to calculate labor productivity of a production line.

Normally, a garment factory captures daily line output. Line output means total number of garments produced by a line. Factory works 8-10 hours a day. To calculate labor productivity of production line, we need these information –

1. Total production volume (output quantity) of the line
2. Total labor worked (headcount) in the line
3. Shift hours and actual working hours in day

Let's say, we have following information of a production line,

- Total production in a day =1200 pieces
- Total labor (operator +helpers) = 37
- Shift time = 10 Hours

Now calculate labor productivity, using the given formula.
Labor productivity = (Total pieces produced/ Total labor input)
= (1200/37) Pieces =32.4 pieces per 10 hours shift

One important note, in this example shift time is 10 hours and we have calculated labor productivity as pieces per 10 hours shift per labor. In case, factory works for 8 hours shift, then with the same manpower, labor productivity of the line will be less as they will produce less than 1200 garments in 8 hours.

Though the above method is used for measuring labor productivity in the garment industry, there is another productivity measure which is known as labor efficiency. This is a comparison of the time spent working productively to the total time spent at work. These metrics are appropriate for analyzing and comparing the productivity of a production line or factory that makes specific apparel products. However, comparing productivity levels across products or operating lines can be difficult because the benchmarks differ from one garment to another.

Calculation of labor efficiency is shown below. Consider the above information for shift time, manpower and daily production. Let's say, SAM (Standard allowed minutes) of the garment is 8.9. Minutes produced by each labor = (32.4 pieces x 8.9) = 288 minutes Available time per worker 600 minutes (10 hours shift per day).

So, Labor efficiency = (Produced minutes/Available minutes) x 100 = (288/600) x 100 = 48%

Practice: Calculate labor productivity for the following cases.

	Line	Product	Daily Production	Manpower	Shift Hours	Estimated production
1	Line-1	Shirt	200	25	8	
2	Line-2	Blouse	250	24	8	
3	Line-3	Blouse	300	30	8	
4	Line-4	Trouser	380	40	8	

How to Calculate Machine Productivity?

Calculating machine productivity is one way of measuring line productivity and factory's productivity. If you are already measuring labor productivity of your sewing lines, you might be also interested in calculating machine productivity for your lines.

Here, I will show you how to calculate machine productivity of a sewing line.

Formula for calculating machine productivity:

> Machine productivity = (Total garments produced by the line/ Total machines used in the sewing line)

Productivity is the ratio of output and input of a process in a defined time frame. Divide garment output of a line by machine input.

Machine productivity calculation method:

To calculate machine productivity of a stitching line, you need following information.
1. Count total machines in a stitching line
2. Count line output at the end of the day (number of garments produced)
3. Shift time (in hours) and total hours factory works

Calculate machine productivity using above formula.

Example: A 40 machines line produced 400 pieces of shirts in 8 hours shift day.

Machine productivity = 400/40 pieces in 8 hours.
= 10 pieces per 8 hours shift

This method is applicable in calculating machine productivity for individual sewing lines as well as the whole factory.

In case your factory shift time is 10 hour, you can measure machine productivity as pieces produced per 10 hours shift. Or You can convert 10 hours productivity into 8 hours machine productivity by using the following this formula.

> Machine productivity in 8 hours = (Total garments produced **x** 8) / (Total machines used **x** Actual shift hours)

Practice: Calculate machine productivity in 8 hours for the following cases.

Line No.	Product Type	No. of machines	Total Pieces Produced (Line output)	Hours worked per Day	Machine Productivity in 8 hours shift
1	Casual shirt	30	400	8	
2	Formal shirt	40	350	10	
3	Trouser	44	420	10	

Answers: (1) 13.3, (2) 7 and (3) 7.63

I hope now you can calculate the machine productivity using the calculation method shown in this section.

It is easier to estimate production capacity of a line for the given products if you have machine productivity data for various product categories.

Difference between SAM Productivity and Production Productivity

In the garment manufacturing industry, manufacturing productivity is measured in terms of labor productivity, machine productivity and value productivity. The labor productivity and machine productivity are presented as number of garment produced (quantity) per labor per shift and per machine per shift, respectively.

Normally, labor productivity and machine productivity are measured for a production line for the selected products.

I see industrial engineers also measure SAM productivity and production productivity. Let me first define these two terms and then I will compare.

SAM productivity

SAM productivity can be defined as total SAM produced per operator per shift. SAM productivity is measuring the manufacturing productivity of a production line or a factory in terms of standard minutes produced per operator.

For an example, let us say in a production line, 40 operators make 400 shirts in a day (8 hours shift time) and SAM of that shirt is 25 minutes.

SAM productivity will be (25x400)/40 minutes= 250 minutes per operator per shift.

Production productivity

Production productivity is the same as the labor productivity and machine productivity. In these cases, productivity is presented in production quantity.

Labor productivity: Number of garments produced per labor per shift.

Consider the above example, in a production line 40 operators and 5 helpers made 400 garments in an 8 hour shift. Total labor count is 40+5 = 45. Labor productivity of that line will be (400/ (40+5)) = 8.89 pieces per labor per shift.

Machine productivity: Number of garment produced per machine per shift.

Let us say, in a line 40 machine is used to make 400 garments in 8 hours shift. Machine productivity of that line will be (400/ 40) = 10 pieces per labor per shift.

Difference between SAM productivity and Production Productivity.

Here are the few points to compare these two productivity terms.

- SAM productivity is presented in **standard minutes produced** irrespective of product SAM. On the other hand, the production productivity is presented as number of garment produced.
- Through SAM productivity you can easily compare performance of two lines or two factories, irrespective of the apparel product types, and style.
  When you use production productivity (labor productivity or machine productivity), comparison of two line's performance can be done when they make the same style and same product category. If two lines make two different products having different SAM, performance cannot be compared with production productivity data.
- You can measure production productivity of a factory or a line without product SAM. But to measure SAM productivity you need to know product SAM.

Garment factories those do not have IE and do not establish standard minutes (SAM) for their styles, cannot calculate SAM productivity. But they can calculate production productivity (machine productivity).

Monthly Efficiency Report

Monthly efficiency is one of the basic and very important information to the management of a company. Efficiency figure shows how a line, a floor or a plant is performing compared to other lines and compared to the factory's benchmark efficiency.

Line efficiency level affects the labor cost of garment. So, each line supervisors are given monthly efficiency target to meet the target labor cost. Factory planning and factory capacity are also measured based on the line efficiency data.

In this article, I will show you how the monthly efficiency report of a sewing line is prepared. And how to calculate monthly efficiency of a sewing floor. To know the monthly efficiency calculation method, read the following steps.

Step 1: Efficiency calculation formula

A standard formula is used to calculate the efficiency of a line. Efficiency is the ratio of total minute produced by line and total minutes available to the line. Efficiency is measured in percentage form.

> Line efficiency% = (Total Standard minutes produced by a line / Total minutes available to the line) x 100

Step 2: Prepare Monthly efficiency calculation table on spreadsheet (Excel Sheet)

To track monthly production data and manpower details, create one table on an Excel sheet as following that includes these information.
- Number of operators (include helpers if manual SAM is added to your Operation Bulletin) working in the line
- Total working hours per day (shift hours and Over Time hours as applied)
- Total garment produced by the line at the end of the day
- Standard Minutes (SAM) of the style that is produced by that line Also add columns for date and styles for reference.

Add three more columns on right side of the table and add formula into the cells of those columns as following
- Total minutes available (No. of operators **x** working hours per day **x** 60)
- Total standard minutes produced (Line output **x** Garment SAM)
- Efficiency% (Total standard minutes /Total available minutes) * 100

Add 31 rows in the table after table header. Add dates from 1st to 31st on the first (date) column.

Table-1: Monthly efficiency report data entry sheet

Line#1								
				Line supervisor: Ganesh				
Date	Style	No. of operators	Working hours	Line output	Garment SAM	Total Minutes available	Total minutes produced	Line Efficiency
1-Nov								
2-Nov								
3-Nov								
4-Nov								
5-Nov								
6-Nov								
7-Nov								
...								
...								
30-Nov								
Total								

Add one more row at the bottom for monthly totals. Now your monthly efficiency calculation sheet is ready. In the Table-1 a blank efficiency report sheet is shown.

Step 3: Data Capturing

At the end of each day fill this sheet date-wise with actual information of manpower, working hours, production quantity in pieces and garment SAM. If factory does overtime, add shift hours accordingly. If factory works on the weekends, fill the data on weekends as well.

When month ends, you will have a sheet filled with 31 days production data. Exclude or keep blank for holidays and weekly off days. A filled monthly efficiency sheet for one line shown in next page (Table-2).

Table-2: A filled table with whole month's production data

	Monthly Efficiency Report (For the Month of Jun 2020)							
Line#1								Line supervisor: Ganesh
Date	Style	No. of operators	Working hours	Line output	Garment SAM	Total Minutes available	Total minutes produced	Line Efficiency
1-Jun	Sty201	50	8	215	25	24000	5375	22.40%
2-Jun	Sty201	50	8	251	25	24000	6275	26.15%
3-Jun	Sty201	50	8	297	25	24000	7425	30.94%
4-Jun	Sty201	50	8	365	25	24000	9125	38.02%
5-Jun	Sty201	50	8	350	25	24000	8750	36.46%
6-Jun	Sty201	49	8	356	25	23520	8900	37.84%
7-Jun	Sty201							
8-Jun	Sty201	49	8	378	25	23520	9450	40.18%
9-Jun	Sty204	49	8	200	24	23520	4800	20.41%
10-Jun	Sty204	49	8	250	24	23520	6000	25.51%
11-Jun	Sty204	49	8	300	24	23520	7200	30.61%
12-Jun	Sty204	48	8	350	24	23040	8400	36.46%
13-Jun	Sty204	49	8	355	24	23520	8520	36.22%
14-Jun	Sty204							
15-Jun	Sty204	50	8	350	24	24000	8400	35.00%
16-Jun	Sty204	49	8	351	24	23520	8424	35.82%
17-Jun	Sty521	49	8	180	26	23520	4680	19.90%
18-Jun	Sty521	49	8	270	26	23520	7020	29.85%
19-Jun	Sty521	48	8	350	26	23040	9100	39.50%
20-Jun	Sty521	50	8	360	26	24000	9360	39.00%
21-Jun	Sty521							
22-Jun	Sty521	47	8	348	26	22560	9048	40.11%
23-Jun	Sty521	47	8	350	26	22560	9100	40.34%
24-Jun	Sty521	48	8	350	26	23040	9100	39.50%
25-Jun	Sty522	48	8	240	25	23040	6000	26.04%
26-Jun	Sty522	50	8	300	25	24000	7500	31.25%
27-Jun	Sty522	50	8	340	25	24000	8500	35.42%
28-Jun	Sty522							
29-Jun	Sty522	49	8	345	25	23520	8625	36.67%
30-Jun	Sty522	50	8	360	25	24000	9000	37.50%
	Total			8161	674	612480	204077	33.32%

Step 4: Monthly efficiency of a line

In the row of monthly totals, you have whole month's data of total minute produced and total minutes available to the line. On the efficiency column you will have monthly line efficiency.

In the above example:
- Total minutes available: 612480 Minutes
- Total standard minutes produced: 204077 Minutes
- So, monthly efficiency of this line is: 33.32 %

How to calculate floor efficiency?

Let us say that you have multiple sewing lines in a floor. You need to calculate average efficiency of a floor. How do you calculate that?

To calculate average line efficiency of a production floor, use the following steps after step 4.

Step 5: Calculate total minutes available to the floor

- Create same spreadsheet for each line as shown in Table-1.
- Fill data line wise. Sum line wise monthly available minutes to calculate total available minutes for a floor.
- Create another report template for line-wise monthly production and efficiency data as shown in the Table-3. You do not need to capture styles for this.

Step 6: Calculate total standard minutes produced by the floor

Sum up line wise standard minutes produced of all sewing lines belong to one floor and calculate total standard minutes produced by the floor.

Step 7: Calculate floor efficiency

Finally, use the efficiency calculation formula to calculate sewing floor efficiency. To calculate average line efficiency of the floor, use total minutes produced by the floor and total available minutes combined of all lines, instead of calculating average of line efficiency data.

For example, I have taken four lines on one floor and line-wise efficiency are as shown in the Table-3.

Table-3: Floor efficiency calculation sheet

Line No.	Total minutes available	Total standard minutes produced	Efficiency %
Line-1	751800	202953	27.00%
Line-2	440400	213220	48.42%
Line-3	498360	180356	36.19%
Line-4	492006	215320	43.76%
Floor Total	2182566	811849	37.20%

I hope you understand the steps for preparing monthly report explained here. If you are not preparing monthly efficiency report, you can start making this monthly efficiency report.

Calculate Line Efficiency for Multiple Styles

Many engineers find difficulty in calculating line efficiency on the day when one style is going to end, and a new style is loaded to the same line. Or when multiple styles are running in a line at the same time.

When we talk about running multiple styles in a line, one can have following situation –

- multiple styles are running in the line at the same time
- the previous style is over with some output and a new style is loaded at the back of the line

Here I will discuss the line efficiency calculation method in both cases. As I have mentioned it earlier, there are two ways for calculating line efficiency.

Formula#1:

> Line Efficiency% = ((Total line output **x** Garment SAM) / (Total operators **x** Shift hours **x** 60)) **x** 100

We write the formula as
= (Total SAM of produced garments / (Total worked hours **x** 60)) **x** 100

Formula#2:

> Line Efficiency% = (Sum of SAM produced by individual operators / Sum of total minutes attended by each operator) **x** 100

We can write the formula as
= (Total SAM produced by the line/Total worked hours **x** 60) **x** 100

Normally garment factories follow formula#1 to calculate line efficiency. Because -
- It is easier to calculate line efficiency,
- Factories do not have the facility to record operator wise output (SAM produced) and
- Factory is concerned about line output rather than the individual operation output.

When two or more styles are running in a line and you need to calculate line efficiency combined of styles, how to use formula#1 to calculate line efficiency? Here is the formula –

Line efficiency = {(Total output of Style-A x SAM of style-A) + (Total output of Style-B x SAM of style-B)} x 100 / (Total operators x shift hours x 60)

> **Example:** Let's say in a line, 20 operators worked for 8 hours and line produces 200 pieces of style-A and 200 pieces of style-B. SAM of style-A is 15 minutes and style-B is 20 minutes. Line efficiency combining two styles output.
> Line efficiency %= ((200 x 15 + 200 x 20) x 100)/ (20 x 8 x 60)
> = (3000+ 4000) x 100/9600
> = (7000/9600) x 100 = 72.9%

Note that though this is widely used methods for line efficiency calculation. With this method, you do not get an accurate efficiency of the line. Therefore, to get accurate line efficiency figure use formula#2.

To calculate line efficiency that runs more than one style you need to measure production of each operator irrespective of styles and measure actual attended minutes by each operator to the work. Then calculate operation wise total produced SAM by multiplying operation SAM to output at each operation.

> **An example:** Let's say, an operator makes two different operations of two different styles. SAM of operation#1 is 0.50 minute and operation#2 is 1 minute and she made 400 pieces for operation#1 and 200 pieces for operation#2. Calculate efficiency of this operator.
> Here, total SAM produced by the operator is equal to 400 minutes (400x0.5 +200x1). Working hours is 8 hours (480 minutes).

In the similar way, calculate SAM produced by each operator and sum up to calculate total SAM produced by the line. Calculate the sum of individual operators working hours. You will get total working hours and total SAM produced by all employees working in the selected line.

Finally, calculate line efficiency% by using formula#2. Efficiency= (400/480) x 100= 83.33%

Method of Calculating Cost per Minute of a Sewing Line

Question: *How we can calculate the cost per minute of a garment sewing line in a very easy way? Show me an easy method of calculating cost per minute of a sewing line. ... asked by an OCS reader.*

Cost per minute of a line is calculated for direct labors. Direct labors like sewing operators, helpers, line supervisor and quality checkers. There are two types of Cost per Minute of a sewing line -
1. Estimated cost per minute of a line and
2. Actual cost per minute of a line

Here we will see the method of calculating both.

1. Estimated cost per minute of a line:

To use this calculation, select one salaried operator. With the monthly salary amount, **salary per minute** is calculated using this formula. (In case a factory gives piece-rate payment to their operators, they can use minimum labor wages of the state/country for calculating factory's cost per minute.)

> Salary per Minute = (Total salary of a line for one day / Total minutes available)

Per minute salary of the direct labor and cost to company to produce one standard minute are different. This can be understood by applying the line efficiency factor. If an employee works 480 minutes in a day and produces 480 standard minutes, then salary per minute and cost per minute will be same. In this situation operator's efficiency is 100%. But in practice very few operators work at that level.

So, we need to incorporate line efficiency parameter to calculate cost per minute of the garment production. Here is the formula for calculating labor cost per minute of a sewing line.

> Labor cost per Minute = (Salary per minute / Line efficiency%)

To calculate the total salary of a line, include all operators and helpers working in a line. You can also include quality checkers and line supervisors and feeder if they are not considered under overhead. So, you need to find the following three parameters to calculate the estimated cost per minute of a line.

- Total salary of a line for one day
- Total minutes available
- Line efficiency%

Where, Total minutes available = (Number of total labors × Daily working hours × 60)

Let's say the total salary of a production line is ₹6000.00, total manpower 20 and line efficiency 40%. Factory works 8 hours a day (480 minutes a day).

Salary per minute = (6000 / (20 × 480)) = ₹ 0.625
Cost per minute = (0.625/40%)) = ₹ 1.56

In this example, we got cost per minute of the line is ₹ 1.56.

2. Actual cost per minute of a line

The formula used to calculate the actual cost per minute of a line = (Actual Total salary for the day/ (Total garment produced × SAM))

So, you need to find the following three parameters -

- Calculate the actual salary of all operators and other direct labors working on the line in a day
- Total pieces made by the line
- Garment SAM of the running order

Let' say, for one day the total actual salary of a production line is ₹6000.00. As the above, let's say 20 operators are working in the and the line has produced 200 garments. The garment SAM is 20 minutes.

Therefore, Actual cost per minute of the line = (6000 / (200 × 20)) = ₹1.50

In this example, line efficiency is (200 × 20/480 × 20) = 41.667%

Now you know the methods of calculating actual cost per minute of the produced garment and projected cost per minute of a line from the given line efficiency data.

Use of Display Boards on the Sewing Floor

An information board also known as display board is used to display production information of the sewing lines, like production status, various performance reports, performance trend charts etc. Information boards are normally hanged on the wall of shop floor or on a pillar.

A display board is made of white board marked with columns and rows or a pin board. Production writer writes production information on this board hourly or at a certain intervals with markers. Multiple lines' production can be displayed in one production board or each line may have separate production board. In the pin board printed reports and charts are displayed.

Most common display board found in the garment manufacturing factories is the hourly production report.

Advantages of keeping a display board on the floor

Visual display is one of the techniques of lean manufacturing. An information board at the end of the sewing line is an application of such lean tool. A display board contains wide range of information. Few major benefits of the display board are listed below.

1. Everybody of the floor can see production status and line performance on the information board. Display board helps managers to monitor line supervisors. Managers do not need to ask supervisors each time they need to know line's output. They can look at the display board and can have all information. Even production managers keep one production board in their cabins.
2. Hourly production reports alarm line supervisors that how far they are behind the target production. Sewing operators also come under pressure when display board shows lower production than the target one.
3. Information associated styles such as style number, buyer name, total order quantity is also displayed in hourly production report.
4. When buyers visit your factory and see display boards with production information, they become confident that they are working with a well-managed company.

Display board information

A display board can include various kind of information. List of information or reports that can be displayed on a whiteboard and on a pinboard are listed below.

- Line wise today's number of sewing operators and helpers working in the line, order no. or style name that is running in the line, name of the Line supervisors and maintenance personnel.
- Line layout with machine name (Printed sheet)
- Today's Hourly line output (production) data
- Loading quantity of cuttings for today and WIP in the line
- Yesterday's line output, efficiency, and productivity report
- Trend chart of the efficiency information of the current month
- Display of KPIs achieved Vs Goal
- End of Line quality performance trend chart
- Displaying top 5 performers of the line or floor with their photograph

Information that can be displayed are not limited to the above list. If your focus is on different parameters, you can display that information on the display board.

KPIs for Garment Manufacturers (Key Performance Matrices)

Key Performance Indicators (KPIs) are measured to assess where the factory currently stands and to find key focus areas where management needs to look at. The top 9 KPIs are listed and explained in this article that are measured by garment manufacturers (export houses) in the apparel industry. Normally, analysis of these KPIs is carried out monthly.

1. Factory Efficiency Percentage

Factory efficiency indicates how efficiently sewing lines are run in a factory. This indicator is important because the capacity planning of the factory and projected garment making cost is measured based on factory efficiency.

For the factory efficiency calculation, you need to include total minutes produced by all lines and total hours attended by direct labors in the sewing floor. Target factory efficiency may vary depending on the order quantity.

For the detailed efficiency calculation method, refer to another article *"How to calculate efficiency of a production line or batch?"* For the factory efficiency – calculate total minutes produced by all lines and total minutes attended by all lines (includes sewing operators and helpers who are doing garment operations).

> Factory efficiency% = (Total minutes produced × 100)/Total minutes attended.

Table-1: Factory efficiency calculation

Line No.	Style	SAM	Prod. Qty	Man-power	Worked Hours	Produced Minutes	Available Minutes	Line Eff%
1	Sty-1	20	410	30	8	8200	14400	**57%**
2	Sty-2	35	300	35	8	10500	16800	**63%**
3	Sty-2	20	410	30	8	8200	14400	**57%**
4	Sty-3	35	300	35	8	10500	16800	**63%**
Factory Efficiency						37400	62400	**60%**

2. Man to Machine Ratio

When it is factory's Man to Machine ratio (MMR), all employees of the factory are considered under manpower. Therefore, Man: Machine =

Total manpower: Total sewing machines available in the factory (machines those are in use).

For example, if a factory has 500 sewing machines and total manpower of the factory is 1100, then man to machine ratio is 1100:500 = 2.2.

This ratio may vary product to product and on the organization structure of a garment factory.

3. Cut to ship ratio

This is a ratio of total cut quantity and total shipped quantity of an order. This indicator is measured order wise and monthly shipped orders. To keep buffer (for damaged, defective garments) factory cuts extra pieces than order quantity. For example, Let's say a factory received an order of 20000 pieces and they cut 20200 pieces (1% extra cutting). They shipped 20000 pieces out of 20200 pieces cut.

Cut : Ship = 20200:20000 = 1.01.

This indicator is measured to control surplus quantity after shipment, reduction in extra cutting and damaged garment. Target Cut to ship ratio is 1. Extra cutting cost extra money to the factory in raw materials and processing cost.

4. Order to ship ratio

The buyer expects to receive full quantity from the supplier on what they have ordered. But many times, it has been observed that suppliers not able to ship the full quantity. This is the most important factor that buyer uses for vendor evaluation. This indicator is calculated as = (Total order quality/ Total shipped quantity). The target of Order to Ship ratio is always 1. It is good if the factory can ship higher than order quantity (only if the buyer accepts extra quantity).

5. On time delivery rate

How much shipment did not meet target shipment date is analyzed at the end of each month? Target on-time delivery of each style is to meet shipment delivery date. If not meet reasons of not meeting delivery date are analyzed. This ratio is calculated as

= Total orders shipped on time/Total orders shipped in the month.

For example, if factory shipped 18 styles on time or before the shipment date, out of total 20 styles shipped in a month then On-time delivery of that month is = (18/20) *100% = 90%

6. Average style changeover time

The time gap between the previous style completion (last piece out from the line) and first piece output of the current style is known as style changeover time. Shorter changeover time is considered as better performance level. It varies style to style and production systems. Time of Change over time of each style is recorded an average changeover time of the factory is measured.

7. Right First Time (RFT) quality

This indicator is represented in percentages. Total audits passed in the first time out of total audit conducted by auditors. Right first-time quality is measured in various stages of garment processing and analysis is done audit wise. Higher value (percentage) of right first-time quality is considered as better performance of the factory.

8. Quality of Production

The quality level of each department is measured in DHU and Percentage defective unit. Higher the value of DHU higher alteration time and higher cost incurred in quality. DHU is measured using the following formula.

= (Total defects found in checking/Total garment checked) x 100

9. Downtime percentage:

Downtime is one of the topmost reasons for less factory efficiency. Factory analyses major downtimes to control and improve machine and operator utilization. The top 5 reasons for downtime (also known as non-productive time, and off-standard time) are line setting, operators sitting idle, no feeding, machine breakdown and no planning for a line.

Garment manufacturing business more specifically garment export business is a profitable business if factory performance is measured and management work to improve performance level step by step. Each of the above KPI plays the role on production cost, factory's reputation, and profit margin.

To learn more about KPI, measuring KPIs and KPI dashboard template read my eBook *"Garment Maker's KPI: Why Measure and How to Measure"*.

KPI Dashboard for the Sewing Factories

KPI dashboard is an effective reporting tool for the production control in the garment industry. In a garment factory, the present reporting system is likely as the following -

You prepare reports on the production status of all departments. In the morning, you email those reports to your managers and take print out of these reports.

And managers are supposed to review all reports. You made a detailed report thinking that your manager should be aware of production status from cutting to finishing, the performance level of each department, each sewing lines and style wise cost per piece, and details of orders those are slipping from the deadline.

Now, ask yourself.

Do they need the detailed reports with tables of thousands of data on each of 10 reports to read line by line? You know, exactly what figures managers review. Show them those data only.

Normally, managers just review key figures. Even if they want to read all the reports, they cannot do it, because it would consume a lot of time. Instead of going through the reports, they would prefer to call you and ask for reasons when they find any issues in the report.

So, why are you wasting your time making long reports and wasting papers for printing those long reports daily morning?

I will show you how to make your production reports effective as well as more interesting. You can make a short report using one of the following methods-

First Method: Instead of 6 to 10 pages reports prepare single page report with important data and numbers.

Second Method: Just make graphs of each analysis and status report and write down key issues that need to be seen urgently by higher management.

Third Method: Prepare report dashboard including all production reports. I call it KPI Dashboard. See the sample KPI dashboard in the following image.

Instead of printing multiple reports make one dashboard and send the dashboard to your managers by e-mail. Such dashboard displays everything that need to be communicated in terms of factory performance.

Industrial Engineer's Digest

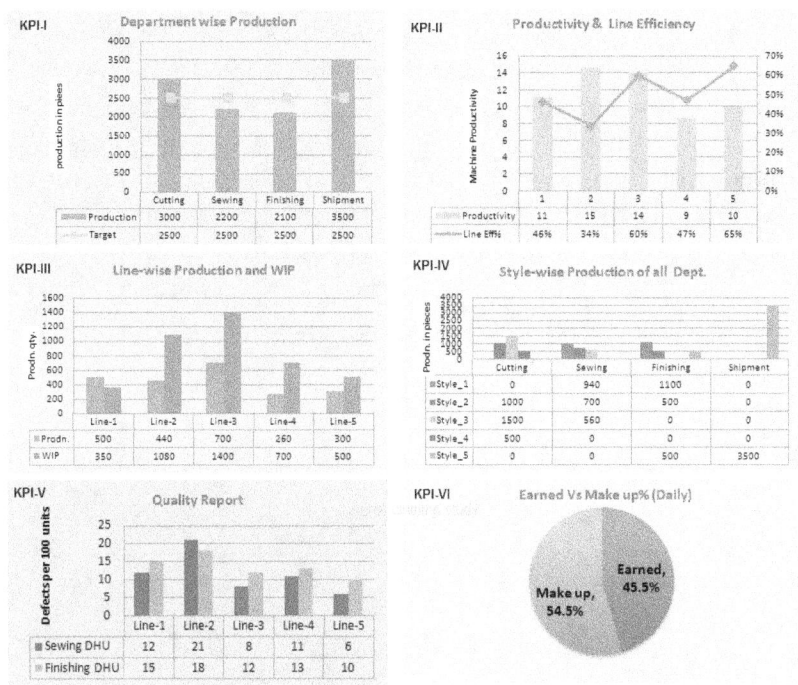

Figure - KPI Dashboard

On the above dashboard, I have included 6 graphs. From these graphs, you can check the production and quality status and factory performance.

How to make this kind of dashboard

I have made one Excel template for data entry and designed graphs on a spreadsheet. I collected data required for following reports and entered data on the template.

1. **KPI-I:** Department wise yesterday's production Target Vs Actual Production
2. **KPI-II:** Line wise machine productivity and efficiency figures of the last production day
3. **KPI-III:** Line wise production and WIP reports
4. **KPI-IV:** Style wise production of all lines
5. **KPI-V:** Quality Report (Sewing and Finishing DHU)

189

6. **KPI-VI:** Earned Vs Make-up% of the sewing operators

Once data is entered in the template, I get all the graphs on the designed on the Excel template. Graphs are then copied in a PowerPoint slide. My dashboard is ready.

The same way you can create graphs for your key reports. Take screenshots of the graphs. Prepare one power point slideshow. Or you can show graphs on the spreadsheet (excel sheet) in a separate worksheet.

A spreadsheet is a better option as you can save all information the same location/file. When someone needs to see reports in detail can open tables for respective reports.

Calculation of Man to Machine Ratio (MMR)

The **man to machine ratio** is an important KPI in the manufacturing sector. The same is applied in apparel manufacturing. In this article, I will explain the term man to machine ratio (MMR), its importance and method of calculating man to machine ratio.

1. **What is the man to machine ratio?**

    Man to Machine ratio is defined as the total workforce in a factory to the total number of operational sewing machines.

    Man to machine ratio widely varies depending on the on the organization structure. If a factory has only production related departments (cutting, sewing, maintenance, quality and IE) then this ratio will be low. On the other hand, if a factory includes manpower from the departments like product design & development, sampling to finishing to shipment and documentation then this ratio will be high.

2. **Why measure MMR?**

    a. By measuring this ratio, factory management assesses how many personnel are employed per machine. This ratio gives a clear indication of indirect cost ratio against the direct labor cost of a company.
    b. Man to machine ratio (MMR) analysis is done to control overhead cost. Every employees and workers cost to a company whatever their salary is. Note that different factories may have different MMR based on the factory size.
    c. Factories those measure this ratio and use it to control cost have a benchmark man to machine ratio. When MMR increases in a specific month, factory checks where manpower increases.
    d. Manpower may increase in sewing floor such as maker man, pressman and helpers or number of staffs. Department heads need to confirm if that additional manpower is required permanently, or requirement is style basis. According to that new MMR is updated.
    e. MMR is measured to check factory's MMR and compare it with the industry benchmark.

3. How to calculate Man to machine ratio?

To measure Man to Machine Ratio (MMR) of a factory -
- Count total number of employees working in the factory (include all manpower).
- Count total number of active sewing machines. Consider only operational sewing machines that are being used in a current month (data analysis month)

For example, let's say in a factory, total 1000 manpower is employed, and the total operational sewing machine count is 500 number. In this case, Man:Machine = 1000:500 = 2:1 (or you can MMR is 2).

In another example, let's say total manpower is 800 and the total operational machine is 500.

Then Man:Machine = 800:500 = 1.6:1 (simply 1.6).

Exercise: Calculate MMR for the following cases shown in the following table.

Factory	Total Staff	Number of Workers	Number of Functional Machines	Total Manpower	MMR
A	40	300	200		
B	50	400	350		

How to Calculate Man to Machine Ratio for a Double Shift Plant?

Normally, the man to machine ratio (MMR) is calculated considering a single shift. Though sometimes factories work overtime for 2-4 hours in a day, I have never seen or heard that factories derive Man to Machine ratio differently for OT hours.

In machine productivity calculation, we consider working hours per shift. Factories used to define their productivity as units produced per 8 hours shift or units produced in 10 hours shift. Factories convert overtime hours into manpower, where 8 hours equal to one manpower.

Normally, garment factories do not work for double shifts. Though it is common that textile mills and textile processing houses work for 24 hours (3 shifts) a day. The double shift means the same set of sewing machines and equipment are utilized for two different shifts in a day. In the second shift, differ group of workers work the shop floor.

I had never calculated man to machine ratio for double shift plant. To answer to this question, I am sharing my views on how man to machine ratio should be calculated for a double shift plant.

When machines are utilized for two shifts, machines should be counted twice for man to machine ratio calculation. Otherwise, if we count machines only once and count total manpower including two shifts' attendance, man to machine ratio will be near to double of the actual MMR ratio, which would be an incorrect presentation of the MMR of a factory. I hope you will agree with me on this.

To make the MMR realistic, I would suggest using the following formula for calculating man to machine ratio where a factory works two shifts in a day.

> Man to Machine ratio for Double shift = (Total manpower of shift-1 + total manpower of shift 2) / (Total machine utilized in first shift + total machines utilized in the second shift).

Let's say total manpower in the day shift is 800 and operational machine is 500. In the night shift manpower is 500 and machine count 400. MMR = (800+500): (500+400) = 1300:900 =1.44

Calculate Number of Days Needed to Complete an Order

Normally, garment factories prepare line loading schedule in advance. The line loading planning is done based on the projected daily production capacity of a line. In this article, we will see the method of calculating the number of days needed for completing an order of 35000 pieces. I have also explained the formula for calculating the production capacity of a line.

The problem and the line performance details are given below.
Order Quantity: 35000 pieces
Daily Shift: 10 Hours
Garment SAM: 26.5 Minutes
Manpower: 125 per line
Number of lines: 4
Line Eff: 50% and Learning curve: Day-1:15%, Day-2: 35% and Day-3:45%, Day-4 onwards:50%
Absenteeism: 5% (Attendance: 95%)

First, let me explain the capacity calculation formula with absenteeism:

> Production Capacity= (Daily Available minutes / Product SAM)

Daily Available minutes = (Daily Shift hours x 60 x Manpower x Line Eff% x Employee attendance %)

In the first day of loading projected production capacity (when line Efficiency =15%)
= (10 x 60 x 125 x 15% x 95%)/26.50
= 403.30 pieces (403 pieces)

In Day-2 production capacity (line efficiency 35%) will be
= (10 x 60 x 125 x 35% x 95%)/26.50
= 941.04 pieces (941 pieces)

In Day-3 production capacity (line efficiency 45%) will be
= (10 x 60 x 125 x 45% x 95%)/26.50
= 1209.91 pieces (1210 pieces)

Day-4 and onward projected line capacity will be (line efficiency 50%)

In Day-4 production capacity
= (10 x 60 x 125 x 50% x 95%)/26.50
= 1344.34 pieces (1344 pieces)

Day-wise total production capacity from 4 lines and cumulative production is shown in the following table. In the last day, lines do not need to work 10 hours for stitch the balance quantity.

Table-1: Daily capacity calculation

Day of working from loading	Line Efficiency	Daily projected output per line (rounded off)	Total Project capacity including 4 lines	Cumulative production
Day-1	15%	403	1612	1612
Day-2	35%	941	3764	5376
Day-3	45%	1210	4840	10216
Day-4	50%	1344	5376	15592
Day-5	50%	1344	5376	20968
Day-6	50%	1344	5376	26344
Day-7	50%	1344	5376	31720
Day-8	50%	820	3280	35000

In the above chart, you can see till Day-7, the total projected production volume is 31720 pieces. To reach 35,000 pieces you need to produce 3280 pieces on 8th day of production. The production needed from each line is 820 pieces.

From the above calculations, 4 lines need to work for 8 days to produce 35,000 pieces of an order.

Calculate WIP in Cutting, Sewing and Finishing Section

Work in progress (WIP) or in-process inventory includes the set at large of unfinished items for products in a production process. These items are not yet completed but either just being fabricated or waiting in a queue for further processing or in a buffer storage.

In garment manufacturing, WIP is generally measured in pieces (number of garments). WIP can be measured as the total pieces loaded in a process minus total pieces come out from that process. The result is the WIP of that process.

If you only want to calculate current WIP level or how many pieces are there as a WIP in a section, use the following formula –

Cutting Room WIP:

When one calculates cutting room WIP, only stock cut pieces is considered as cutting WIP. Fabric those are layered on table or received from fabric store for cutting is not considered as WIP.

> Cutting WIP = (Total cut quantity - Total quantity sent to sewing section)

For example, cutting department cuts total 5000 pieces for a Style-A and 3000 pieces have been issued to sewing department. Then cutting room WIP will be 2000 pieces.

Sewing Section WIP:

WIP level of sewing section can be calculated for a line, for a floor or WIP between two operations. You need to use specific formula for what you are calculating WIP level.

> Sewing Line WIP = (Total pieces loaded to the line - pieces completed)

Let's say, till date you have loaded 3000 pieces to the sewing Line-1 for Style-A and total 2500 pieces are out from the line including today's production. Sewing line WIP for the style is (3000-2500) = 500 pieces.

Finishing Room WIP:

> Finishing room WIP = (Total pieces received from sewing section - Total pieces packed)

Let's say, 2500 pieces are received from the sewing section for style A. Finishing section has finished total 1800 pieces and has sent the finished quantity to the packing section. Finishing room WIP would be 700 pieces (2500 – 1800) pieces.

I hope now you can calculate department wise (process wise) WIP quantity. Follow the above calculation method to measure WIP between two consecutive processes.

You can make WIP level of all these processes in a single report. Here is an example.

Line	Style	Order Qty.	Total Cutting	Total pieces loaded to sewing section	Cutting WIP	Total pieces come out from the line	Sewing section WIP	Stitched garment sent to finishing	Total garment finishing	Finishing WIP
1	Style-A	5000	5000	3000	2000	2500	500	2500	1800	700
2	Style-B									
3	Style-C									
4	Style-D									
5	Style-E									

In the next article, I have shared 7 ways for reducing WIP in bottleneck operations in a sewing line.

7 Ways to Reduce WIP from Bottleneck Operations

A bottleneck operation limits the production of a sewing line.

In a bottleneck operation, garments are piled up when the other operations get dry in the line. From the bottleneck operations line gets lowest number of garments.

Efficiency of the line and labor productivity are affected by work in process (WIP) level at bottleneck operations.

By reducing WIP level at bottleneck operations labor productivity can be improved. In this article, 7 practical ways are discussed that will help you to reduce WIP from the bottleneck operations.

1. Method improvement:

For the method improvement, you first need to conduct method study of the bottleneck operations. Method study is the in-depth monitoring and analysis of the way an operator performs her task. Once the method study is done, look for better ways to save time and less effort by the operator. This could be done by improving work motions or providing work aids like gauges, folders, attachments, trolleys, movers, tracks, or machine automation. As much time as you save through method improvisation from bottleneck operation, you get better production that will in turn reduce WIP level.

2. Share workers' capacity:

In the sewing line, you will easily find some operations with higher capacity than the capacity required for line's target production. Take few pieces from the bottleneck operation to the nearby operation which has potentially higher capacity. Maintain this capacity sharing in a certain interval. If you do not have performance based (individual or group) incentive scheme in that case normally operators don't like to do others work. Operators who had potential capacity to share his/her capacity for bottleneck operation intentionally reduce his/her speed to balance the bundle completion time with the line speed.

However, if you like to use this method, you must be careful on the operator with which work is shared has same type of machine as that of bottleneck operation and operator has right skill required for it.

You can also utilize spare capacity of adjacent line's extra capacity operations.

3. Add additional manpower or machine:

This is the easiest way to increase production at bottleneck operations. I would suggest do not go for this option. Instead, first try to improve working method.

In case you find the potential capacity of the operator is too less than the target output of the line, then consider for adding extra machine (and an additional operator). Calculate each machine capacity and give them target for two machines. According to it add additional sewing machine.

By following this you can drastically reduce WIP level. It must be kept in mind that your machine productivity may reduce in some cases when you increase number of sewing machines in a line.

4. Improve workstation layout:

The first and last movement (i.e., pick up garment parts and dispose it) performed by an operator in an operation cycle depends on the workstation layout i.e., where you place the cuttings and where the disposal bin has been placed.

A closer look to workstation layout of the bottleneck operators will help you to find out whether the layout is following principles of best workstation design mentioned by Hiba in his book *"Improving Working Conditions and Productivity in the Garment Industry: An Action Manual"*

- o Position material, tool and controls within easy reach
- o Use jigs and other devices to save time and efforts

This could be relating to ergonomic issues like light, fan, seat adjustment, etc. Redesign workstation to reduce material handling time and get increased production.

5. Better operator allocation:

Each operator has different set of skill level (operations they generally perform) and efficiency at work. So, at the time of operator selection, do it correctly as much as you can. Allocate high content jobs to highly skilled (matching to the job) operators. And for low skilled operators select jobs that required low skill to perform and that has comparatively low work content. In this step, you need to check whether operator selection was done correctly or not.

For example, if you have operator grading system then check if any 'A' grade operator is allocated on a low skilled and high capacity operation, and B/C grade operator to the bottleneck operations. If this be the case both the operators could be swapped to get more production out of bottleneck.

6. Work for extra hours:

If you find additional machine set up is not possible due unavailability of machine or space and above steps don't make much difference in WIP reduction, then follow this step. You can work overtime at bottleneck operations (if your factory policy permits for working extra hours) to make enough pieces and create WIP for the following operations.

7. Use time saving tricks:

A lot of time is spent by operators in material handling and associated jobs. Here I have listed few time-saving tricks. These may be helpful to you.

- If the bottleneck operation is having sub-parts of operations, they can be bifurcated and given to helpers or other nearby operators. In this way operator can produce extra pieces and keep smooth flow to the line.
- Use UBT machines for thread trimming instead of hand clippers or trimmers.
- Check previous operation if there is any quality issue due to which the operator is not receiving proper input to work on. This may reduce production and create bottleneck operation.
- The operator could be provided with extra bobbins and needles (depends on company needle policy) to save the winding time and requesting for new needle and receiving it.

These steps are applicable in any garment factory. You can apply these tips and reduce WIP from bottleneck operations in the production lines.

Why Measure WIP in a Garment Factory?

Work in progress (WIP) includes the set at large of unfinished items for products in a production process. These items are not yet completed but either just being stitched or waiting in a queue for further processing or in a buffer storage. WIP is in-process inventory.

In garment manufacturing, WIP is normally measured in number of garments. The end-of-day WIP of a process is the number of total pieces loaded today in the process plus the previous day's WIP minus total pieces come out from that process. In a garment factory, production WIP data is measured every day by the production department or by the industrial engineering department.

Why measure WIP?

Maintaining a minimum level of WIP at each sewing workstation is necessary to have continuous work for workers. But neither a high WIP nor the low WIP is good for a production line.

Here are a few reasons for which a factory needs to measure the production WIP.

- To know whether a line (or workstation) has high WIP level or low WIP level compared to the targeted/planned WIP
- If a line supervisor doesn't look at operations where WIP is piling up, the production line will become imbalance soon.
- Allowing unnecessary WIP in the line would increase the inventory requirement as well as financial involvement in a period.
- To keep the production line running, it is necessary to feed the production lines and workstations continuously. If the workstation does not get enough work, it will be on the idle situation and no production would be generated from that workstation. In such cases, a line would loss potential output from that workstation.
- Most cases, a high WIP is the main source of many issues like quality issues, low productivity, and unnecessary delay in production.

By measuring the WIP, the factory can plan for their next day's work requirement. In the next article, I will show you how to prepare WIP report of a garment factory.

How to Make WIP Report in Garment Production?

As part of MIS reports, I developed WIP reports for a few garment factories. In this article, I will share one case of making WIP reports.

The factory used to make high fashion dresses for the international market. To make such kind of products factory needs to involve a lot of manpower in hand-repairing process. Inside the repair section, there are multiple sub-processes. Hand-repair, checking of the repaired pieces, alteration of the hand-repaired piece, checking of the altered-hand-repaired pieces.

In such production environment, the factory required process-wise and sub-process wise WIP report data for managing daily production and keeping the cost per garment under control.

This factory keeps a manual record of each garment- who has repaired the piece, at what time the piece was issued to a repairman and when the repair was completed by the repairman. But from that manual record, they do not get WIP data at different processes at any moment.

So, I was asked for make one WIP report for them and replace the manual data recording of each piece. While discussing with the factory manager, I experienced and learned new things. I will be sharing those here, in this post.

How do you prepare the production report in your factory? I can understand that you are not making the same kind of products. Still, you might be making WIP report in your factory.

The fact is you cannot ignore the WIP report. Whether you want it or not, you need to make it. WIP is the work that you need to finish and WIP is money that lying on the floor.

Production WIP

It means the volume of work that is under process in the production processes. If it is sewing line WIP, it means the total number of semi-stitched (unfinished) garments lying in the line. Garment cuttings are loaded but not yet come out from the line. Garment pieces lied inside the line, spread over the workstations of the line.

In an earlier article, _Calculating WIP level in cutting, sewing and finishing section,_ I have shown WIP calculation formula as:
- Cutting WIP = (Total cut qty - Total qty sent to sewing),
- Sewing Line WIP = (Total pieces loaded to the line - total Pieces completed)

- Finishing room WIP = (Total garment received from sewing - Total pieces finished)

Technically above calculation method is correct. But in this method timeline is not defined. Total pieces loaded to the line since when? Is it today or week-to-date or week-to-month loading quantity? If you consider it week-to-date total pieces loading quantity, then at the start of the week you must count previous week's balance WIP. You need to complete garments and make zero WIP in the line. But that is not possible to clear out the line in every week or in every month. Then how to measure the WIP in garment production processes? Read the below explanation and formula.

WIP calculation formula

WIP of a sewing line need to be calculated using the following formula:

> Sewing line WIP = (Total garments loaded today + Previous day's balance WIP - Today line output)

Steps for calculating WIP

Step-1: Prepare an Excel template. You can follow the template shown in Figure-1.
Step-2: Add WIP calculation formula on the Excel Spreadsheet
Step-3: Collect required information for WIP report. You need to record 3 data on the selected date -
   (1) Today's loading quantity, (2) Today's output quantity, (3) Previous day's WIP quantity (closing balance).
Step-4: Enter the manually collected data in your spreadsheet template. You will get the complete report as per your report format.

The same way you can prepare WIP for other sections. Like cutting section WIP, Finishing section WIP.

The production recorders can easily keep previous day's WIP records in their daybook and carry it forward as the opening balance for today. This closing balance carries forward and becomes the opening balance for everyday WIP calculations.

The physical registers are okay for recording and maintaining day to day data. But not enough for data analysis and information sharing with management. WIP report should be generated automatically and should be available for viewing it at any moment you need it.

WIP Report

To generate this report, I used their SQL database. The factory had production data from the real-time production tracking system. See the below example of such a WIP report.

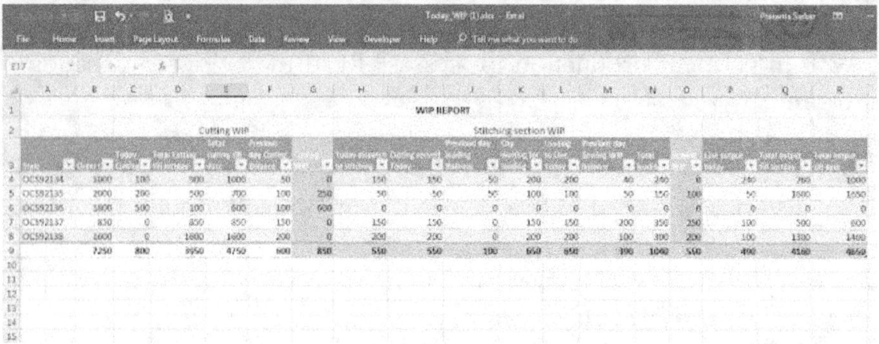

Figure-1: WIP report format showing cutting and stitching section data

As the report headers are not readable in the above figure, I am listing header have in the following table.

A	B	C	D	E	F	G	H	I
Style number	Order quantity	Today cut qty.	Total cutting till last day	Total cutting till date	Previous day's cutting balance	Cutting WIP	Today cutting sent to sewing section	Cutting received today by sewing
J	K	L	M	N	O	P	Q	R
Previous day's loading balance	Qty. waiting for loading	Today loading qty.	Previous day Sewing WIP balance	Total Issue to the line	Sewing WIP	Line output today	Total output till last day	Total output till date

In the above WIP report, cutting section WIP and sewing section WIP are included. The WIP is showing till-date total production quantity which is optional for the WIP report.

Note: Excess WIP on the shop floor at any process is not good for many reasons. If you find bottleneck operation, act on reducing WIP from the bottleneck operations.

Reporting and Data Analysis of Overtime Work

When a factory works more than the regular shift hours in a day, the extra hours are considered as overtime work. In India, 8 hours per day excluding the break time and 6 days week are considered the normal working hours. When a worker works more than 8 hours in a day or more than 48 hours in a week, those additional hours are counted as overtime working hour (OT hours).

For an example, in a garment manufacturing factory, Normal working hours: 9:00 a.m. to 5:30 p.m. (30 minutes lunch break is included).

Overtime working hour starts after 5:30 p.m. If a factory works till 7:30 p.m., daily overtime work will be 2 hours.

Before starting the overtime work, some factories give 10-15 minutes break. Including the break-time shift is extended up to 7:45 pm. The factory shift start time, lunch break duration, and shift close time may vary from one factory to another.

As per the Factories Act 1948, every adult (a person who has completed 18 years of age) cannot work for more than 48 hours in a week and not more than 9 hours in a day. According to Section 51 of the Act, the spread over should not exceed 10-1/2 hours. (source: paychek.in).

I have written this article to guide one engineer, who asked me this question. *What MIS report should I prepare for the overtime working hours, so that management can understand and clear? I want to show our management that overtime working is not beneficial for the factory.*

Let me come to the question. MIS reports prepared for overtime working hour and data analysis made for the KPIs, effected by the overtime working hours. Overtime hours simply called as OT hours. You can do the following data analysis. In the data analysis, I have taken hypothetical data for showing you examples how you can do the comparative analysis.

Overtime working report and data analysis

The first and foremost thing you should do is prepare a detailed report of the overtime hours worked in your factory. Whether the overtime was planned one or a sudden need for overtime work, include it in your overtime report. The report can be simply captured in a register by the production writer. Or the industrial engineering department can keep daily overtime hours record. In practice, factories prepare a list of workers who will be doing overtime and the list is signed and approved by the factory manager (or by an authorized personnel). I had observed this practice in many factories. If you are following the same OT approval method, overtime hour calculation will be easier.

By capturing the data, prepare a detailed report, weekly overtime hours, monthly overtime hours and percentage of overtime compared to regular working hours.

Table: OT Hours Report

Date	Line Number	Style number	Regular shift hours	OT hours	Number of workers @ regular hours	Number of workers @ OT hours	Garment SAM	Production in Regular hours	Production in OT hours
20-08-2018	L01	ABC012	8	2	40	30	25	500	60
20-08-2018	L02	ABC012	8	2	38	32	25	450	50
20-08-2018	L03	ABC012	8	2	40	35	25	400	45
20-08-2018	L04	ABC012	8	2	40	20	25	425	50
20-08-2018	L05	ABC012	8	2	40	26	25	460	50
20-08-2018	L06	ABC012	8	2	36	36	25	510	55

To get more insights of the factory overtime working hours and its impact on the garment cost, I would suggest you capture additional information for regular hours and OT hours as listed below (if you are not capturing this information), like

1. Line wise daily production,
2. Daily working hours,
3. Daily manpower,
4. Daily produced SAH and
5. Daily line efficiency for regular hours, OT hours and the whole day efficiency.

Start capturing such data and keep for data analysis. A sample data capturing format including this information are shown in the below image (Fig-1). You can use the same format for preparing your overtime report.

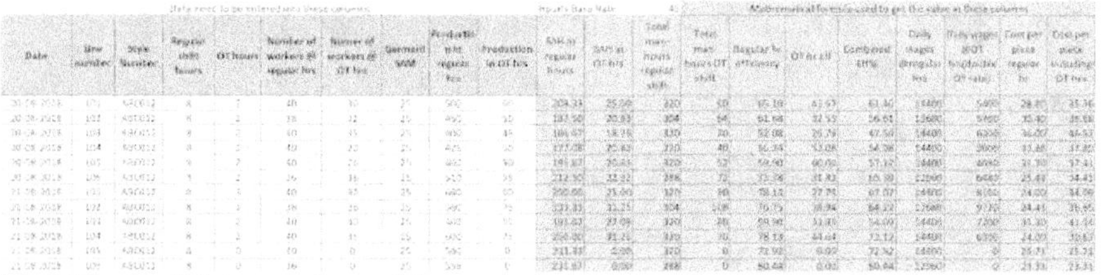

[Table image: detailed data capturing and analysis report]

Fig-1: The detail data capturing and data analysis report template

You can further analyze the stitching quality data of the regular hour work and overtime work hours.

List down the reason for overtime working

When a factory is doing overtime, you should know why overtime is required. Are you doing OT just to meet the daily production target, or you need OT to complete the production of a certain order as it is getting delayed due to some uncertain reasons? Whatever it is, make a note for analysis. From my observation, I have listed a few reasons for doing OT.

- Non-availability of cuttings and fabrics in the day and production need to be completed by the date.
- Non-availability of the trims
- Overbooking of orders
- Shortage of special machines and machine breakdown
- Low productivity
- Stitching quality issues
- Change of production plan
- Need for completing a delayed order

If it is a regular practice of doing 2 hours of overtime, then you need to improve your planning. Most cases workers slow down their production rate in the regular working hours, so that factory work for overtime. By doing this an employee can earn extra money on daily basis.

Comparative analysis of cost per pieces with overtime and without overtime work:

Production cost per piece is one of the production KPIs. Calculate production cost in both scenarios - when you are doing over time and paying the double payment for overtime hours and paying money for food to employee and staff. If your factory does not pay double overtime, consider single payment for calculating extra expenses.

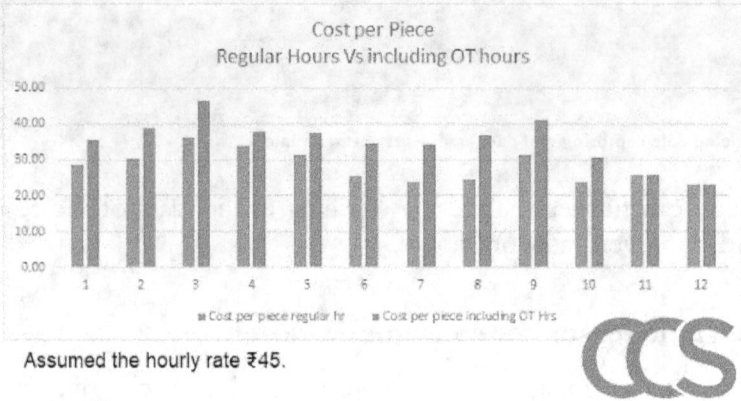

Assumed the hourly rate ₹45.

Fig-2: Cost per piece comparison

This chart (Fig-2) will show your management, that overtime work cost to the factory. As the cost per piece increase when you do overtime work. Note, for some cases overtime work can be considered, to avoid penalty of late shipment. That overtime should be pre-planned.

Comparative analysis of the line productivity with overtime work and without overtime work

As mentioned above capture daily working hour - regular hours and overtime hours, daily production - in regular hours and in overtime hours. Prepare comparative analysis report using these data.

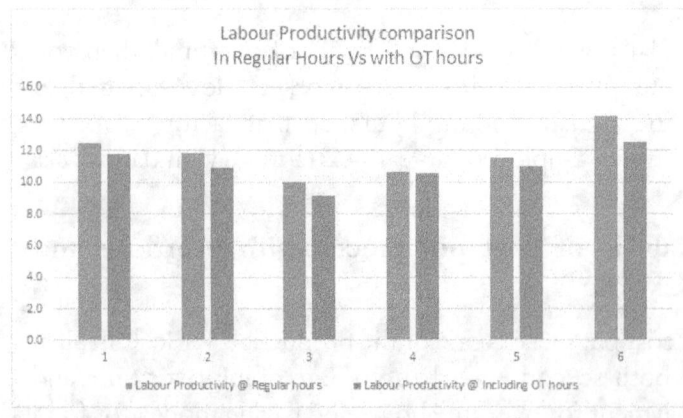

Fig-3: Labor productivity comparison

In Fig-3, you can see average labor productivity reduction of 6 lines is 7% when the factory does the overtime for two hours a day. The above comparative analysis shows that labor productivity goes down when you do excess overtime.

Line Efficiency data comparison

Compare the line efficiency in the regular hours and line efficiency including OT hours production. See an example of the same in Fig-4.

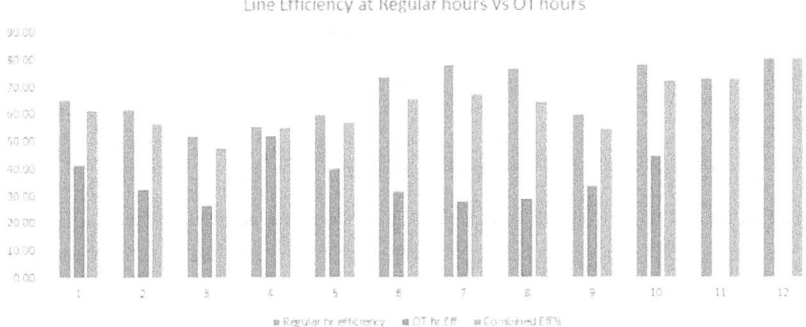

Fig-4: Line efficiency data comparison

Negative Impacts of excessive overtime work

Overtime can be beneficial for both employees and companies. It provides the company with the flexibility to cover unexpected absences and changes in demand without hiring more staff and it gives employees extra income at a premium rate. However, overtime has its downsides too. While many employees will happily take as much overtime as is available, there is growing scientific evidence that relying too much on overtime can lead to numerous problems for an operation. (circadian.com)

The negative impacts of too much overtime to the employees and to a factory are listed in the five categories.

- Increased health problems
- Increased safety risk
- Decreased productivity
- Increased absenteeism
- Increased worker turnover Rates

Management would not like to be the reason for the above problem for the employees. The above statements are given considering the non-garment manufacturing industry. So, the last two statements may differ when we are talking about the garment industry. Many workers prefer working in a factory where they get a chance of earning more money. Overtime working is considered one of the ways of earning extra money for them. But the first 3 statements will imply to garment factory employees.

If you look at the factory compliance, regular working per week is 48 hours. If you are working more than 48 hours in a week you are not abiding by the compliance rule. So, you may need to hide your actual working hours in your reporting and payroll. This is considered an unethical practice.

How to reduce and eliminate overtime?

If you are in favor of not doing overtime, as you know there are many negative impacts to the workers, you need to find out alternative ways that help you meet your daily production target, avoiding urgency of production completion, and increasing employees' daily earning. Prepare a note on how you will be managing all these three challenges after eliminating over time.

1. Increase daily earning:

Daily earning can be increased by providing performance incentive and attendance bonus. Incentive bonus scheme can be designed for an individual operator as well as for a group. In the group incentive system, all operators working in the same line and making the same style as a group will be eligible for incentive earning if their line output meets the target production qty. In individual incentive bonus system, whether the line meets the target, an operator can earn incentive if their individual efficiency reaching the target efficiency level.

By providing an attendance bonus to the employee, a factory can gain many other benefits in daily production and line balancing. You can proof this by using your actual data.

2. Improve Labor Productivity

Daily production target can be reached by improving labor productivity. Productivity can be increased by introducing a performance incentive scheme and managing the production line well. There are many reasons for the low productivity. If you are an industrial engineer, you might be already knowing why daily production is low in your factory.

You need to list down the possible ways that can be implemented on the shop floor based on your order and present status. Prepare a report for before and after implementation result and show it to your management. They will understand it is possible to meet the daily production target by working on regular shift hours.

3. Manage Production Delay

Thinking of how to **manage the production delays**. A good planning, production scheduling and execution of the schedule are required to avoid the delay. Order execution should be done from sample approval, fabric sourcing and order planning. It is often found that in a factory, due to the communication gap and coordination issue between departments, things are getting delayed. You can use the production scheduling and production execution tool for managing orders and reduce the delay in process completion.

Order Completion Report

Order completion report, mostly known as OCR, is an MIS report that is made once an order is shipped to customer. Processing of an order is not complete until OCR is made and submitted to the cost accounting department. Profit from an order after shipment of goods can be assessed only when OCR is made.

Normally, the production planning department makes this report. But management can assign this responsibility to any other department for making this MIS report. It does not matter who make the report. Important point is that correct and complete information must be collected from departments those involved in making the order.

The order completion report is compiled together to analyze figures on following parameters. The objective of this report is to measure

1. Number of pieces are shipped against the order quantity. Primary target is to ship 100% of order quantity. But in case buyer agrees to accept extra quantity with same price, manufacturers try to ship extra pieces up to buyer's acceptance limit. In case factory shipped less quantity that ordered, buyer may charge for penalty.
2. Number of garments are shipped against cut quantity. Garment manufacturers, normally, cut extra unit of garments than the order quantity as there is a chance of producing defective garments. Extra cutting means excess cost incurred in raw materials and manufacturing.
3. Whether shipment is delivered on time or it is delayed.
4. How many garments are rejected and damaged?
5. Stock level of raw materials after shipment: Stock of fabrics and other raw material are measured. Planned average of fabric and trims consumption is compared with actual consumption used in the order.
6. Planned production cost Vs Actual production cost
7. Total Dollars factory has earned after completing an order.

Calculated data of the above parameters determine performance level of the company. Company sets its own benchmark or target level for each performance criteria. Data tells us how factory manages its inventory, quality of the product etc. On-time delivery and full order quantity delivery is very good sign for good and long-term relationship with buyers.

Based on the OCR data, departments those are responsible for mistakes or lowering the factory performance, learn and try to correct in the following orders. For example, factory may end up with lot of fabric inventory after shipment due to wrong estimation consumption. This reports help employees to learn from their mistakes.

A sample report of details of Cut to Ship Quantity of an order is shown here for the reference. For each of the above criteria data is collected in the specific format.

Table: Sample format for order completion report

	Order Tracking Sheet (Details of cut to ship quantity)											
	Order number	Shipped date	Order Quantity	Total Cut Quantity	Extra cutting @ 3%	Cutting issued to Sewing Section	Stitched quantity	Garments damaged in Stitching	Pieces issued to finishing department	Finished Quantity	Rejected/Damaged garments in finishing	Packed quantity	Shipped quantity	Excess /Surplus quantity
1	#OCS235	5th Jan	10000	10300	300	10300	10251	49	10251	10187	64	10187	10100	87
2	#OCS456	10th Jan	2000	2060	60	2060	2050	10	2050	2032	18	2032	2000	32
3														
4														

You can further maintain the surplus/excess garment quantity by color and by size.

Useful Formula for IEs and Production Team

In this book, you have come across many calculations and formula used in calculating useful measures. In the section, I have listed some of those mostly used formula for your quick reference.

As these are common calculations in apparel production, most of the factories use Excel templates for preparing such reports and calculations where the excel template are pre-filled with formula. We just follow the report template. We do not need to calculate every time we make a report. So, many times we are not clear about how to calculate a specific measure.

1. Hourly production target

Hourly production target calculation is a part of an OB and daily production report. Hourly production target is calculated for individual operators as well as for a line as needed. In an OB, production target is calculated for each operations. Following formula is used to calculate hourly target from one machine.

> Hourly production target = (60 / Operation SAM)

Note: This target is calculated at 100% efficiency. In case you want to plan hourly production target at certain percentage, multiply the above by line efficiency or multiply by individual operator's efficiency as per your need.

In the Table-1, few examples are shown for calculating hourly production target at 100% efficiency.

Table-1: Examples of hourly target calculation

Sl. No.	Operation	SAM	Hourly Target (60/SAM)
1	operation 1	0.78	76.92
2	operation 2	0.85	70.59
3	operation 3	1.2	50.00
4	operation 4	0.9	66.67

2. Daily Line Target

Like the hourly production target, daily production target of a line is calculated. This target is given to line supervisors. Based on the daily production target production planning and total production time needed to complete an order is calculated. Following formula is used to calculate daily line target.

> Daily Line Target = (Shift hours × 60 × No. of operators working in a line × Line efficiency %) / Garment SAM

See the example in the following table.

Table-2: Daily production target calculation

Line no.	Manpower	Shift hours	Average line efficiency%	Garment SAM	Daily production Target (pieces)
1	40	8	65%	30	416.0
2	40	8	60%	24	480.0
3	42	8	62%	42	297.6
4	46	8	58%	38	337.0

3. Individual operator efficiency

We measure individual operator's performance (efficiency %) for operator grading, skill matrix, and daily employee performance reporting. In case you are distributing performance-based incentives to individual operators, you need to measure individual operator efficiency too. This is the formula to calculate individual operator efficiency.

> Individual Operator Efficiency% = (Total minutes produced × 100/Total minutes worked)

Here,

Total minutes produced = (Total units produced × operation SAM)

See the following examples (Table 3) for calculating individual operator efficiency. In case an operator is doing more than one operation, first calculate total minutes produced in each operation and sum up total minutes produced in all operations by an employee.

Table-3: Examples of individual operator efficiency calculation

Sl. No.	Operators	Operation	SAM	Unit produced	Hours worked	Efficiency%
1	Operator-1	Operation1	0.78	500	8	81%
2	Operator-2	Operation 2	0.85	480	8	85%
3	Operator-3	Operation 3	1.2	300	8	75%
4	Operator-4	Operation 4	0.9	480	8	90%

4. Line Efficiency

Line efficiency is the common measure for measuring line performance and factory performance. Performance of a production line vary depending on operators' skills, product design, line balancing, presence of lost time, style run time and many other factors. A better efficiency means higher output and reduced production cost. So, we aim to improve line efficiency as much as possible. Line efficiency is calculated using following formula.

> Line Efficiency% = (Total minutes produced x 100) / (Total hours worked x 60)

Where, Total minutes produced by the line = (Line output x Garment SAM)

Total hours worked by the line = (Number of operators x Shift hours)

Table-4: Examples of line efficiency calculation

Sl. No.	Line	Style	SAM	Units produced	No. of manpower	Hours worked	Eff%
1	Line-1	Style-1	24	400	25	8	80%
2	Line-2	Style-2	30	480	35	8	86%
3	Line-3	Style-3	12	300	20	8	38%
4	Line-4	Style-4	20	500	35	8	60%

Note: Include helpers and workers doing manual operations in the manpower, in case SAM of the manual operations are included in garment SAM.

The above-mentioned formula is majorly used for line efficiency calculation. This is known as output-based line efficiency. Line efficiency can be also measured in a different way. Instead of line output data, individual operators produced minutes and hours worked are calculated first and later their produced minutes and hours worked are summed up to calculated line's produced minutes and total hours worked. To many, this is not easy to calculate individual operators' output, their total produced minutes, and actual hours they worked in a day. The second method is only practical if a factory is using real-time production tracking system.

5. Machine Productivity

Machine productivity is measured to know the average production per machine per day or in the defined time frame. A line with 48 machines producing 480 shirts and another line with 36 machines producing 400 shirts (same design). How do you figure out which line doing better? This can be answered simply by measuring machine productivity.

Productivity is the ratio of output and input. Machine productivity of a production line is calculated using following formula.

> Machine Productivity= (Line output / No. of machine used)

Machine productivity is measured as production per machine per shift day. See the following example (Table-5).

Table-5: Example of machine productivity calculation

Sl. No.	Line	Machines used	Production	Machine productivity
1	Line-1	48	480	10
2	Line-2	36	450	12.50
3	Line-3	36	420	11.67

6. Labor Productivity

Productivity is also measured as labor productivity. Instead of machine input in this case line output is divided by labor input to calculate labor productivity.

> Labor Productivity = Line output / No. of total manpower (operators + helpers)

7. Standard Time

Standard time is the time allowed to an operator to carry out the specified task under specified condition and defined level of performance. This is a standard definition for standard time. Some additional time is added to basic time to arrive standard time of a task.

In practice none can work throughout the day without taking rest. Operators need time for relaxation from fatigue. Various allowances are relaxation allowance, contingency allowance (like machine breakdown) and bundle allowance (for PBS system).

> Standard Time = (Observed time × observed rating) + Allowances

Allowances includes relaxation allowance, contingency allowance, bundle allowance and machine allowances.

The basic constituents of standard time are shown in the following chart (Chart 1). This chart shows how standard time is made up from observed time and basic time of a job.

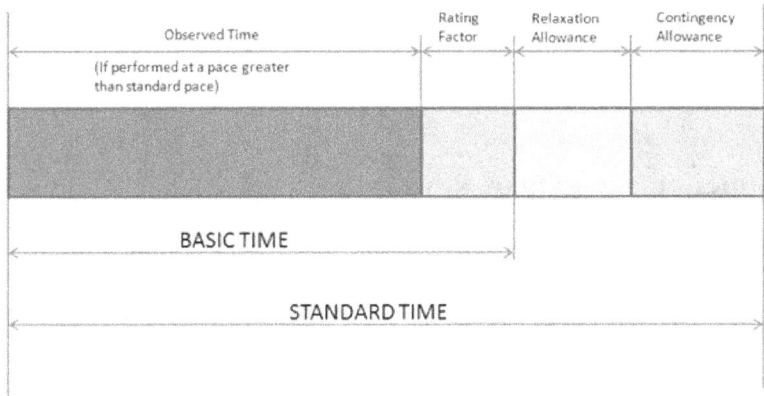

Chart-1: This chart shows how standard time is made up. Source: Introduction to work study by ILO

8. Machine utilization percentage

Machines and equipment are primary resources of garment manufacturing unit. These resources must be utilized as much as possible to improve factory performance. But due to work unavailability or less work, some machines (like specialized machines) are partially utilized by the factory. In such case, same machine is shared for multiple lines. Machine utilization is calculated using below formula.

> Machine utilization% = (Actual machine running time x 100) / Time available

If you are running one of your special machines only 4 hours in an 8 hours shift day, utilization of that specific machine would be 50%.

9. Labor cost per unit

Labor cost is part of garment FOB. For every order we have a target labor cost. We need to track the actual labor cost and control it. So, we measure it after a style is loaded.

Here is the formula to calculate labor cost per garment.

> Labor cost per unit = (Total cost incurred in labor wages / No. of garments produced) in a day

10. Production capacity of a line

Production capacity is measured in available minutes as well as in number of pieces. First, we will see how available minutes in a line are calculated. Then we will calculate capacity in number of pieces.

> Line capacity per day= {(No. of machine x Dily work hours x 60) x (1 – absenteeism %)} x Efficiency % (capacity in minutes)

Production capacity (in pieces) = (Available minutes / Garment SAM)

Table-6: Example of production capacity calculation

Line No.	No. of Machines	Minutes/Day (Daily working Hrs. X 60)	Line Efficiency	Absenteeism %	Capacity available (in minutes)	Garment SAM	Production capacity (pieces)
1	30	480	60%	10%	7776.00	15	518.40
2	28	480	72%	8%	8902.66	20	445.13
3	32	480	65%	9%	9085.44	20	454.27
4	32	480	55%	10%	7603.20	20	380.16

Like other garment manufacturing units, you might be also measuring some key production measures to keep track of production performance and controlling the production. Whatever measures and formula you are using must be standardized and approved by the authorized person internally. Everyone in an organization should be aware how the specific measures are calculated.

It is good to have a poster of commonly used formulas and keep the poster in your workspace. A poster would be helpful for learning for newcomers.

Part-III

How to Guides

How to Calculate Garment SAM?

Standard allowed minute (SAM) is used to measure the work content of a garment. To calculate the garment SAM, first you need to establish the standard time of the individual operations of a garment.

In this article, you will learn how to establish standard minute of individual sewing operations.

For estimating the cost of making a garment, SAM value plays a very important role. In the past, scientists and apparel technicians did research on how much time to be allowed to do a job when one follows the standard method of doing the job. According to the research study, minute value has been defined for each movement (motion) needed to accomplish a job. Synthetic data is available for each movement and micromotions.

There are other methods through which one can establish SAM of a garment without using synthetic data.

Method - 1: Calculate Garment SAM Using Synthetic Data

When we talk about synthetic data, we refer to MTM (Methods-Time Measurement) data. There are few Predetermined Motion Time Systems (PMTS) that provides synthetic data. Garment factories buy the PMTS system and use it to establish 'Standard Time' of garment items and other sewn products. Follow these steps to estimate standard time of an operation.

Step 1: Select one operation for which you want to establish standard time (SAM).

Step 2: Study the motions of that operation. Stand by the side of an operator (experienced one) and observe the operator how he is doing it. Motion study can be done through videography.

Note all movements used by the operator in doing one complete operation cycle. See carefully again and recheck your note if all movements and motions are captured and correct. (for example, motions are like - pick up parts one hand or two hands, align part on the table or machine foot, realign plies, etc.)

Step 3: List down all motion sequentially. Refer to the synthetic data for TMU (Time measuring unit) values. For synthetic data, you can refer

MTM codes and TMU of different motions and TMU values). Sewing Performance Data table (SPD) is a free database that can be used for establishing standard time.

Now, you got the total TMU value for one operation. For example, let's say it is 400 TMU.

Convert total TMU into minutes (1 TMU=0.0006 minute). This is called as Basic Time in minutes. In this example, basic time is 0.24 minutes (400 x 0.0006).

Step 4: Calculate SAM by adding bundle allowance and personal fatigue allowance to the basic time. Bundle allowances need to be added as per the bundle size.

> Standard allowed minutes (SAM) = (Basic minutes + Bundle allowances + Machine and personal fatigue allowances)

For an example, add bundle allowances (10%) and machine and personal allowances (20%) to the basic time. After adding these allowances, you got Standard Time (SAM).

Standard Minutes = (0.24+0.024+0.048) = 0.31 minutes.

In this article, I have briefly shared the method of establishing SAM using synthetic data. You need to arrange a PMTS database from one of the software company or you can use basic codes and TMU. The most important part is learning the motion analysis and motion study. In the next page, I have shown one example of time and motion study sheet.

Read article *Secret Behind Calculation of Machine Time in SAM* for better understanding of SAM calculation using synthetic data.

DataS   2018-11-11 03:57:00

Method description
R_OP_S1

SGP00510   Shirt Collar   Topstitch collar
LOCKST 1 Needle   Topstitch collar, 5mm, close neckline.
Shirt M   Shirt Collar

No.	Element code	Element description	Additional text	Frequency	TST
1		Bundle handling			
2	-_2-_BB00ST1	1 Small panel, tied, barcode sticker		10 /	0.040
3		Left side			
4	G80B	Pick up piece and position under foot accurately, 16 - 30 cm		1 *	0.033
5	P79A	Position for side topstitch 0-15 cm		1 *	0.029
6	MA00	Sew, cm: 3, stitches/cm: 4.5, RPM: 4000		1 *	0.010
7	H45	Turn small piece needle down 0-15 cm		1 *	0.011
8	MA4540	Sew, 4.5 stitches/cm, 4000 rpm, cm: 8		1 *	0.016
9		Middle side			
10	H40B	Turn piece needle down 16-30 cm		1 *	0.023
11	H15A	Move or straighten piece without grasp 0 - 15 cm		1 *	0.009
12	MA4540	Sew, 4.5 stitches/cm, 4000 rpm, cm: 3		1 *	0.010
13	P65B	Regrasp with hand movements 16-30 cm		1 *	0.007
14	MA4540	Sew, 4.5 stitches/cm, 4000 rpm, cm: 34		1 *	0.045
15	P65A	Regrasp with hand movements 0-15 cm		1 *	0.005
16	MA4540	Sew, 4.5 stitches/cm, 4000 rpm, cm: 4		1 *	0.012
17		Right side			
18	H45	Turn small piece needle down 0-15 cm		1 *	0.011
19	P65A	Regrasp with hand movements 0-15 cm		1 *	0.005
20	MA4540	Sew, 4.5 stitches/cm, 4000 rpm, cm: 8		1 *	0.016
21		Neckline			
22	H40B	Turn piece needle down 16-30 cm		1 *	0.023
23	P65A	Regrasp with hand movements 0-15 cm		1 *	0.005
24	MA4540	Sew, 4.5 stitches/cm, 4000 rpm, cm: 25		1 *	0.035
25	P60	Regrasp with finger movements		1 *	0.004
26	MA00	Sew, cm: 12, stitches/cm: 4.5, RPM: 4000		1 *	0.021
27	H70	Cut thread with foot motion		1 *	0.005
28	D10D	Dispose, 46 - 80 cm		1 *	0.026
				TST	0.403
				Allowance	1.15
				SAM	0.464

Username: DataS   © timeSSD v01.08.02.27 by DataS   Page 1 of 1
                                         www.timessd.com

Figure: Example of synthetic data for establishing operation SAM | Shared this with timeSSD permission.

Method - 2: Calculate Garment SAM Through Time Study

The second method for establishing an operation SAM is Time Study method. In an earlier article, I have explained the Time study method. We will use that steps for establishing SAM of an operation. Prior to using this method, you must learn time study procedure. The prerequisites of time study method are time study format, stopwatch, a pen/pencil and time study board.

Step 1: Select one operation for which you want to calculate SAM and select an operator who is doing the operation.

Step 2: Time Study: Take one stopwatch. Stand by the side of the operator. Capture cycle time for that operation. (Cycle time – the total time taken to do all works needed to complete one operation cycle, i.e. time from pick-up part of the first piece to next pick-up of the next piece).

Conduct **time study** for consecutive five cycles. Discard the reading if you found abnormal time in any cycle. For better accuracy, you can study more cycles.

Calculate the average time of the 5 operation cycles that you recorded. The time you got from time study is called cycle time. To convert this cycle time into the basic time, you need to multiply cycle time with operator performance rating. Let's say, average observed cycle time is 0.60 minutes.

Step 3: Rate your operator: Rate the operator at what performance level operator is doing the job seeing his movement and working speed. Suppose that operator performance rating is 80%.

Step 4: Calculate Basic Time: In this step, you need to convert the cycle time into basic time. Use this formula.

Basic Time = (Cycle Time **x** Performance Rating)
Therefore,
Basic time = (0.60 **x** 80%) = 0.48 minutes

Step 5: Calculate SAM using the following formula. Add allowance to basic time.

> Standard allowed minutes (SAM) = (Basic minutes + Bundle allowances + Machine and personal fatigue allowances)

Add bundle allowances (10%) and machine and personal fatigue allowances (20%) to basic time

Bundle allowance = 0.48 x 10% = 0.048 minutes
Machine allowance = 0.48 x 20% = 0.096 minutes.
SAM = (0.48 + 0.048 + 0.096) = 0.624 minutes.

An example of SAM calculation is shown in the following table.

	Particulars		Time
A	Observed Time		50 seconds
B	Performance Rating	80%	
C	Basic time	(50 x 80%)	40 seconds
D	Machine allowance (%)	7%	2.8 seconds
E	Bundle Allowance (%)	5%	2 seconds
F	Personal Fatigue (%)	8%	3.2 seconds
	Total Time (C+D+E+F)	(40+2.8+2+3.2)	48 Seconds
	SAM		**0.80 Minutes**

Bundle allowances need to be added as per bundle size. If your bundle size is 10 pieces, divide bundle allowance by 10 for adding bundle allowance to each operation cycle time.

Following the above procedure, we normally establish garment operation SAM through Time Study.

After reading this article, I hope you understood how garment SAM is calculated by industrial engineers. Before you use one of the above two methods, you need to learn **Time Study** method and **Time and Motion Study** procedure.

How to Estimate Sewing Time by Using Machine RPM?

Can you estimate how much time should be taken by a sewing operator to sew a specific length of seam? Here I have explained a method for estimating sewing time from machine RPM.

Machine RPM (revolution per minute) decides how much (minimum) time is required to sew a specific length of seam with specific Stitch per Inch (SPI). I am only considering the machine cycle time of an operation. I use the following formula for calculation.

> Machine cycle time = (SPI x Seam length in inch)/Machine RPM

To use this formula, you need machine SPI, machine rpm, and seam length. SPI specifies number of stitches (needle dips) in each inch of the seam and machine motor RPM specifies how many times a needle dips through fabric per minutes.

A specific seam length gives you how many stitches required to sew complete seam. Use simple mathematics to find machine time required to sew a certain length of fabric.

Machine RPM 2000 (let's take a lock stitch sewing machine), required stitch per inch is 10 and you need to sew 50 inches straight seam (suppose a superimposed seam). So, to sew 50 inches length of a seam total stitches (needle dips) needed = 50 x 10 stitches = 500 stitches

As per machine RPM, machine can make 2000 stitches per minute at its maximum speed. So, to make 500 stitches, machine will need only 15 seconds (500 x 60/2000). I am assuming that no stopping during stitching complete 50 inches seam. Now you got the idea. If you think your machine is running at 1000 RPM, then in this case machine time will be 30 seconds.

You may think that when sewing a garment operator can't make it in a single burst. The above formula gives you clear information about the minimum time is needed to machine time if we ignore all other factors.

How to Reduce Standard Time (SAM) of a Style?

You already know what the **standard time** is. You might be also knowing how the standard time of a garment style is calculated using PMTS data or through time study method. You have standard time for all the products and styles you are making in your factory.

Now the question is how to reduce Standard Time (SAM) of a Garment Style?

I do not need to mention the importance of reducing the standard time of your styles for a garment factory. After the reduction of the existing garment SAM, you will get a new standard time for the same product. The reduced standard time will result in reduced direct labor cost and increased labor productivity.

I am considering that you are already using sewing machines attachments, folders and guides and you are also using the correct standard time. Even after that, there are two ways of reducing the standard time of a garment.

1. By using an automated sewing machine.

Install an automated sewing machine (where applicable and automation available). For making trousers and shirts, there ae automatic machines for making some of the garment components.

2. Through method improvement

Standard Time is measured combining motion times (TMU) of all motions in an operation cycle and allowances. If you can eliminate some motions from the operation cycle, the standard time of the operation will reduce.

By using a PMTS software, you can identify unnecessary motions in an operation cycle and eliminate those. This way you can reduce SAM of a garment style. Later, train your sewing operator to follow the best methods and motion sequence.

If you can reduce the standard time of some of the operations in a style, standard time of the garment will decrease.

How is the Sewing Time Calculated (Machine Time in SAM)?

Standard Allowed Minute (SAM) of an operation is the sum of 3 different parameters, i.e., machine time, material handling time (with personal allowances) and bundle time. Material handling and bundle time is calculated through motion analysis.

Do you know how the machine time is being calculated?

A standard formula is used to calculate machine time. In this article, it has been explained in detail. You will get information on how machine time is calculated for SAM.

Machine time consist of 5 parameters as following
1. Straight machine time (A)
2. Number of stops/starts in a seam (B)
3. Stopping Accuracy (C)
4. Difficulty of material handling (D)
5. Machine allowance (E)

1. Straight machine Time (A):

Straight machine time is explained as the time needed in sewing a seam without considering accurate stop at the end of the seam, material handling and machine allowance. To calculate straight machine time you need Machine RPM, SPI and Seam length.

Machine Time (A) = (SPI × Seam Length)/Machine RPM

To represent machine time in TMU (Time measuring unit), the above value (A) is divided by 0.0006

2. Number of stops/starts in a seam (B):

Every time operator stops machine to align, realign panels or for other motions, a fixed time is added to machine time. Normally it is 17 TMU (for stop 8.5 TMU and for start 8.5 TMU). If machine is stopped twice in a single seam, then 2 times 17 will be added to machine time. Let's represent number stop/start in an operation as 'B'.

3. Stopping Accuracy (C):

At the time of sewing a seam how would you like to stop the machine? Machine stop could be a precise stop or a normal stop. For example, during stitching of a pointed collar, machine must be stopped at the collar point accurately, otherwise you will not get a sharp corner. Stopping accuracy can be categorized in different level of accuracy. A differential TMU value is added to machine time according to stopping accuracy. Let's represent it as 'C'.

4. Difficulty of handling (D):

At the time of sewing garments operator need to hold and guide fabric panels to get stitches on the desired line. Difficulty of handling garment components depends on no. of fabric plies to be sewn, types of fabric is used, shape and type of seam. It is obvious that sewing a straight line is easier than sewing a curvy line. Handling multiple plies is more difficult than single ply. According to the material handling difficulty level the sum of above three (A+B+C) is multiplied with difficulty factor. Factors used such as 1.1, 1.2, 1.3 etc. Let's represent it as 'D'.

5. Machine Allowance (E):

A certain percentage of allowance is added to machine time for machine allowance. Machine allowance is added due to certain non-avoidable tasks such as threading of machine, changing of bobbin and machine breakdown needed to perform during sewing. There is pre-defined allowance percentage for machine allowances. Machine Allowance is added to the result of above four parameters.

To calculate the total machine time, use the following formula.
= [{(straight machine Time + Number of start/stops + Stopping accuracy) x difficulty factor in Material Handling} + {(A+17B+C) x D} x Machine allowance %]
= [{(A+17B+C) x D} + {(A+17B+C) x D} x E%]
= (1+E %) x {(A+17B+C) x D}

An example is shown in the following table.

	Particulars	Parameters
A	SPI=10, seam length=20 inches, machine rpm=3000	111.11 TMU
B	Number of start/stop in the operation cycle	5
C	Stopping accuracy = Normal (0)/Accurate (10)	0
D	Difficulty in handling	1.2
E	Machine allowance =15%	15%
	Machine time (calculated using the above formula) in TMU	270.63

Machine time in minutes = 0.162 minutes

Measure SMV Improvement Percentage

In the garment industry, the industrial engineering department does a lot of improvement work in a factory, like improving productivity, reducing cost per garment, setting up SOPs for various processes. The improvements they do, are measured through various KPIs. But many engineers are not yet measuring some important KPIs. One such KPI is SMV improvement percentage. The SMV reduction or SMV improvement percentage is also considered a KPI for the work study and industrial engineering department.

I will show you the importance of measuring SMV improvement and the formula for calculating this KPI.

Importance of measuring SMV improvement

While industrial engineers aim for the production cost reduction, it is often done by reducing the standard time of the operation. The SMV can be reduced by applying workplace engineering, method improvement, and introducing machine attachment or guides.

- It is required to calculate monetary saving after introducing new SMV
- You must measure all your improvement works. It will help one get noticed by managers.
- You must keep the record for the previous method and the new method with respective operation's SMV. (if the improvement is brought by improving method).
- You can consider this improvement as a continuous improvement project (Kaizen)
- Easier to keep track of operation wise and product wise SMV improvement.

Calculate SMV Improvement Percentage

Let us measure SMV improvement in your factory. This formula is used for calculating SMV improvement.

> SMV improvement percentage = ((Old SMV - New SMV) x 100)/Old SMV

Example: Let us say, old SMV of a sewing operation was 0.85 minute. After improving it, you got a new SMV 0.75 minute. Therefore, SMV improvement will be

= ((0.85- 0.75)/0.85) x 100
= 11.76%

If you are not yet measuring it, you can initiate measuring SMV improvement percentage. In turn you can measure how much money is saved by reducing the operation/garment SAM.

How to Estimate Daily Garment Production?

Production estimation is one of the basic requirements for production management. The production line supervisors need to estimate garment production on daily. Production supervisors, engineers who are working in production, must know how to estimate daily production in a garment factory.

I will explain the calculation procedure of estimating daily garment production from the garment SAM and available working hours.

Production is the total number of garments stitched by operators in a line in a day. Production is also termed as daily line output. To estimate production from a line, we need the following information.
1. Standard allowed minute (SAM) of the garment (the style you will be making)
2. Number of operators working in the line
3. Number of hours line will work in a day
4. Average Line efficiency

We will use the following formula for calculating estimated production.

Formula: The formula for production estimation

> Daily production = (Total available man-minutes in a day/Garment SAM) x Average Line efficiency

Total available man-minutes = (Total number of operators x Working hours in a day x 60)

Example: Let's assume one new style will be loaded in a production line. Style SAM, manpower, line efficiency and daily shift time is as the following.
- Garment SAM: 20 minutes,
- Number of operators working in the line: 30
- Shift hours in a day: 8 hours
- Average line efficiency (Planned line efficiency) =50%

Total available man-minutes = 30 x (8 x 60) = 14,400 Minutes
Therefore, daily estimated production = (14400 /20) x 50% = 360 Pieces

You can expect 360 pieces output from this line if the line efficiency matches as expected (50%).

Daily actual production quantity from a line is directly proportional to the line efficiency; the number of operators and working hours. And production is reverse proportional to the garment SAM. If the efficiency of a line increases, you can expect higher production. Similarly, if the style SAM reduces, you can expect higher output.

In the above calculation, I have not considered these two factors – employee absenteeism and loss time. When you include these two factors, your production estimation will be more accurate. If you have employee absenteeism percentage data and loss-time percentage, you can consider these. Available minutes will be calculated as following.

> Available man-minutes = ((Total number of operators x Daily Shift x 60) (1-Absenteeism%))-Total expected loss-time in minutes

Practice: Using the above formula, calculate daily garment production with the information given in the following table.

	Manpower	Daily shift	Garment SAM	Line Eff %	Absenteeism %	Loss-time (minutes)	Estimated production
1	30	8	25	75	5	120	
2	35	8	24	75	6	150	
3	36	8	30	60	5	200	
4	20	10	15	65	5	160	

Factors that hamper production:

There are few more factors that can reduce production of a line. So, when you calculate daily estimated line output, consider parameters, like machine breakdown, feeding status of cuttings, quality problems, low performing operators, and operator absenteeism.

Answer Key: 1. 407, 2. 489, 3. 324, 4. 487

How to Calculate Production Target and Worker's Bonus in the Initial Days of Production Start?

Question: *Many of my orders are of 5,000- 22,000 pieces per style. My sewing line's output is around 1,200 pieces per day. So, styles are finished within 1 -2 weeks and need to change the style. In the first week of loading style, operators are not familiar with the style and they learn in this period. So, the time measurement is not correct to be a reference. Then in second-week reference can be measured accurately but almost the styles get finished.*

How can I set the target and bonus more adequate in this situation? The bonus amount should be attractive to workers and fair to the factory. Workers work on an hourly wage.

The problem raised in this question is a fact of garment manufacturing factories. When a new style is loaded to a line, operators learn on the initial days and produce less garments. That is one of the reasons for low production in initial days compared to pick production or average daily production.

I will explain how you can set the production target and operator bonus effectively even on the initial days of the production start.

Setting Production Target

You might already know the formula that is used to calculate the daily production target of a style. I have shown the formula here.

> Production Target per day = (Number of operators x Working hours per day x 60 x Line Efficiency%) / (Product SAM)

There are two problems in using this formula calculating production target on the learning stage.
- **You would not know the actual line efficiency** – as efficiency will build up day by day up to pick efficiency level. And without past data you cannot assess what would be the efficiency of day 1, day 2 or day 3.

- **Product SAM** – I am assuming that you are not using Predetermined Motion Time System (PMTS) for measuring product SAM. Production capacity is estimated based on time study of the production line. This is indicated in the question.

The only solution of this problem is that you need to develop database for line efficiency at the learning stage. This period is called as learning curve.

Develop learning curve of your lines (style wise, order wise) by studying styles. Measure the trend - how production build up happens and reach to pick production day by day at the learning stage.

Once you have enough data on efficiency build up (learning curve), you can plan target production for the future orders. Learning curve may vary depending on order quantity and style difficulty level.

Learning curve (Efficiency% Vs Days) of a typical garment factory for one style is shown in the following Figure. In this learning curve, day wise efficiency build-up is shown.

Similarly develop database for standard time for your styles by using Time Study method.

Now you have both data to calculate production target even at the learning stage.

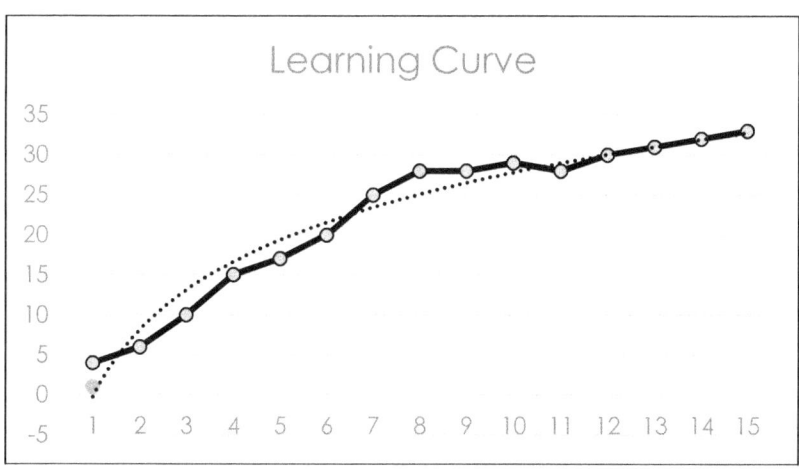

Figure-1: Learning curve

How to set worker bonus?

You can setup the worker's performance bonus at the learning stage for an individual worker or/and for the group without much difficulty.

It is a similar method that you use calculating bonus amount on the normal production days. The difference is - in this period the target efficiency% for becoming eligible for earning bonus to be kept lower than normal production days.

Design day-wise target efficiency to be eligible for earning bonus based on the learning curve. You can keep the bonus amount same per point increment of efficiency. Or slightly more to motivate your operators working harder.

How to Calculate Target Production per Hour from Cycle Time

Question: *Can you please tell me how to calculate pieces per hour of an operator? Let's say average cycle time is 30 seconds. How to calculate the hourly production of the operator?*

The production capacity from the cycle time is calculated using this formula.

Production capacity per hour (in pieces) = (3600/Operation cycle time in seconds)

Note: 1 hour = 3600 seconds

You need to add allowances to the average cycle time to estimate the production capacity realistic. As you know, when you do cycle timing of an operator, she would have been working for her bundle and stitched few garments (operations cycle) one after another without taking a break.

In the cycle time record, the allowances like machine breakdown, threading needle, bobbin change, bundle opening and tying the bundle, personal needs, are not included. Combining all add 20% allowance on the average cycle time.

In the example, the operation cycle time is 30 seconds. After adding 20% allowances, the calculated cycle will be 36 seconds. This is derived from (30 seconds X 1.20) = 36 seconds

Therefore, estimated production quantity per hour = (3600/36) = 100 pieces

Following this method, you can calculate hourly production target for an operation from the operation cycle time.

Read the article titled *How to calculate hourly production target from SAM?*

Industrial Engineer's Digest

How to Balance a Traditional Sewing line?

You know that several operations are involved in making a garment. In mass production, generally a team works on an assembly line and each operator does one operation and passes the garment to another operator to do next operation. In this way, a garment reaches to end of the line. In the assembly line, after some time of the line setting, it is found that at garments are piled-up and in few operations, operators sit idle due to unavailability of work.

When this situation happens in the line, it is called an imbalanced line. Normally, it happens due to two main reasons –

- variation in work content (time needed to do an operation) in different operations
- operator's performance level. To meet the production target, maintaining smooth workflow in the line is very important. So, it is very important to know basics of quick line balancing.

How to balance an imbalanced line has been explained here.

The main objective of line balancing is to eliminate or reduce WIP (work in process) from the bottleneck operations. To do that you need to know which operations are bottlenecks in the line. Through capacity study and target setting, you can find existing bottlenecks in a line.

Tools required for capacity study - stopwatch and spreadsheet and calculator. Follow the below steps to balance a production line.

Step 1: Capacity study:

List down all operations (with operator name) as per operation sequence in a paper. Using stopwatch, cycle time (time study) each operation for five consecutive cycles. With average cycle time calculate the hourly capacity of the operators. (e.g. operation cycle time 30 seconds and total allowances are 20% then capacity is 100 pieces per hour). Draw a line graph with per hour capacity data of all operations performed in a line.

Step 2: Target setting:

With the above capacity data set your target output per hour from one line. Normally, it is calculated using following formula (Target per hour= (Total no of operators **x** 60) /garment SAM). Check current hourly operator production report. Draw a straight line with target output data on the line graph.

Step 3: Identification of bottleneck areas:

Now go to the capacity study table and compare each operator's capacity with the target capacity. Each individual operator whose capacity is less than the target output is bottleneck operation for the line. It is impossible to improve imbalanced line's output without improving the output of the bottleneck operations. A bottleneck operation is like a weak link for a chain.

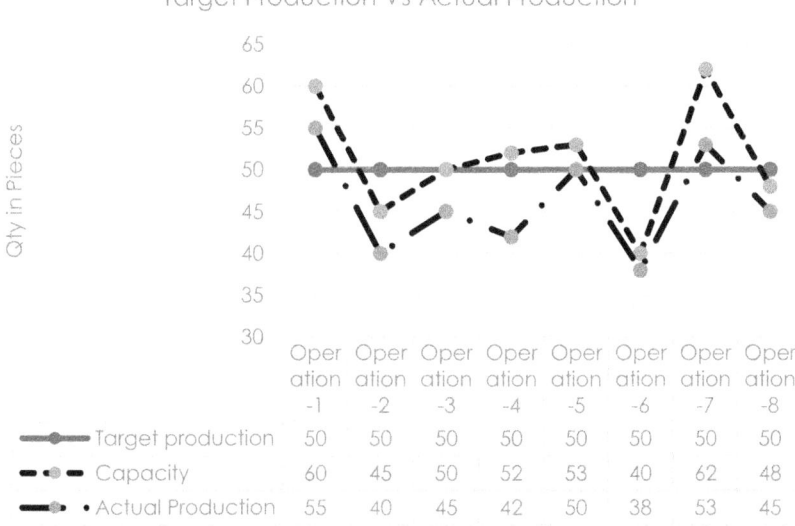

Step 4: Eliminate bottlenecks from the line:

Now to eliminate bottleneck areas, use the following methods which suit best to your situation but don't jump without trying initial steps.

1. **Club operations where possible.** Where there is higher capacity than the target output, give that operator another operation with less work content. Considering machine type and sewing thread colors. (See the Operation-7 in the above chart, production capacity in that operation is 62 pieces but hourly target qty is 50 pieces).
2. **Shuffle operators.** Operations that have low work content, use a low performing operator. And where work content is higher, use high performers.
3. **Reduce cycle time using work aids and attachments.** To assist the operator in handling parts during sewing, positioning cutting

and disposing of the finished task, work aids, guides or attachment can be used. Think of that. If possible, provide the operator work-aids. It will reduce operation cycle time.
4. **Improve workstation layout and improve methods**. Most important area for improving output from an operation is using best workstation layout and best method of work. There is always a chance that through improving the method of bottleneck operations you can do line balancing.
5. **Add more operators to bottleneck operations**. Adding one additional machine is the easy task than others. Before adding one more machine, compare the cost-benefits of putting an additional machine into the line. It can be compared by estimating machine productivity in both cases of before and after adding a machine.
6. **Do extra work at bottleneck operations**. In lunchbreak and tea-break, when each operator of the line goes for a break, bottleneck operator can continue work to feed next to his operator. Later he can take the break. At the end of the day tell this operator to work for one hour extra to reduce the WIP. This step cannot be applicable in many garment factories. Because after the shift, factory does not allow workers for working extra hours.

Important things to keep in mind:

Conduct Time Study hourly basis and check the output of each operator. Once you have eliminated one bottleneck, you will find a new bottleneck operation. Follow the same methodology to eliminate next bottleneck operation and improve line balancing. Line balancing is continuous process.

In the next article, you will learn the line balancing method using operator's skill history.

How to do Line balancing using Operator Skill History?

In the last article, I discussed how the traditional assembly is balanced after line setting and while the line is running. Here, I have explained the second method of production line balancing using operators' skill matrix.

The prerequisite of this method is the sewing operators' **skill matrix**. A skill matrix is the database where you get individual operator's efficiency level, types of operations they can perform, types of sewing machine they can run and operation-wise efficiency. Skill matrix can be developed in the Excel spreadsheet.

Normally, at the time of line setting, operators are selected based on their experience on operations. The calculated skill level of the operators on the operations is not considered in most of the garment factories.

As a result, after a couple of hours, highly skilled operators complete their bundles faster and wait for the work (next bundle). On the other hand, the low-skilled operators stuck with their work. The production rate of low-skilled operators is less than the line's average production rate. Thus, line becomes imbalanced, and a lot of productive time is lost as operators sit idle. To utilize operator's maximum capacity, work allocation must be done based on operator's potential performance level (efficiency) and work must be shared with operators who have excess capacity.

For setting up a balanced line from day one, select operators for the operations based on operator's skill levels and capacity required for the task.

Assume that you already have skill matrix of your operators. For example, a line produces T-shirt and skill inventory of the operators are as following. Operator's skill levels in various operations have been shown in the following Table-1.

Work sharing between the operators may vary depending on production systems. I will explain line balancing method considering that we are balancing an assembly line (Progressive Bundle system). Assumptions:

- Operators will work on a single workstation.
- Work will be shared, or operator's excess capacity will be utilized only where machine type matched, and the operator

has performed that operation earlier (refer to operator skill inventory)
- No time delay when an operator switches from one operation to another.

Table-1: Operator Skill Matrix

	Operator Name	Mark, neck trim & run stitch	Join shoulder	Insert neck trim	Serge margin	T/S on Neck trim	Binding shoulder to shoulder	Attach sleeve	Sew side seam with label	Sew sleeve hem	Sew bottom hem
1	Radha	81	76		75						
2	Seema	65	93	71					90		
3	Anil				80	75	82				
4	Geeta			76		65					
5	Leela			74			86				
6	Roma							77	92		
7	Nirmal					75			65	72	
8	Sunita							80	73	69	
9	Kamal									82	77
10	Sivani									85	82

Step 1: Keep a copy of the skill matrix of all operators working in the sewing line in which a new style will be loaded. You need to develop operator skill matrix if you do not have one as shown in Table-1.

Step 2: Collect Operation bulletin (OB) for the style to be loaded or a style that is running in the line. Operation bulletin must have information such as operation name, SMV of the operations, type of machine is used and hourly target from the line. Refer to following OB (Table-2). Daily target output is 675 pieces in 8 hours shift and hourly target is 85 pieces.

Step 3: Prepare a table on a spreadsheet with the headers as shown in the Table-3. Enter information to this sheet from your operation bulletin. Columns A, B, C and E to be filled from OB.

Step 4: Other columns and formula used for calculation are explained in the following paragraph. Insert formula in your spreadsheet as explained here.

Table-2: A Basic Operation Bulletin (For a T-shirt for example only)

Sample Operation Bulletin					
Style	OCS120402	Description		Crew neck Tee	
Qty:	5000	Shift		8	
Buyer	XYZ	Machine SAM		5.50	
Target	85	Manual SAM		0.00	
S. NO.	Operation	M/C Type	SAM	TGT @ 100%	Calculated manpower	Allocated m/c
1	Make Neck rib & Run stitch	SNLS	0.55	109	1.00	1
2	Join shoulders	4TOL	0.45	133	0.82	1
3	Insert neck trim	SNLS	0.45	133	0.82	1
4	Serge margin	4TOL	0.31	194	0.56	1
5	T/S on neck trim	3TFL	0.34	176	0.62	1
6	Bind shoulder to shoulder	DNCS	0.54	111	0.98	1
7	Attach Sleeves	4TOL	0.78	77	1.42	2
8	Sew side seam with labels	4TOL	0.84	71	1.53	2
9	Sew sleeve hem	3TFL	0.68	88	1.24	2
10	Sew bottom hem	3TFL	0.56	107	1.02	1
	Total		5.50		10	13

Table-3: Line balancing calculation table

A	B	C	D	E	F	G	H	I	J	K	L
Operation	Output needed per Hour	SAM	Output Expected @ 100%	Machines	Operator Name	Potential performance (Eff%)	Hourly Output	Time Available	Balance Qty. (Over/Under)	Time /needed	Spare time
1 Make neck rib & run stitch	85	0.55	109.1	SNLS	Radha	81	88	1	3	0.96	0.04
2 Join shoulders	85	0.45	133.3	4TOL	Seema	93	124	1	39	0.69	0.31
3 Insert Neck trim	85	0.45	133.3	SNLS	Geeta	71	95	1	10	0.90	0.10
4 Serge Margin	85	0.31	193.5	4TOL	Anil	80	155	1	70	0.55	0.45

Output Needed per Hour (B): Target output per hour in pieces. Initially, target output in each operation is 85 and when an operator worked in an operation produce part quantity, balance quantity will be the needed quantity for the second operator doing the same operation.

Standard output (D): Potential production capacity at 100% efficiency. D=(60/ SAM of an operation)

Operator (F): Select the most suitable operator from the skill matrix and write his/her name in the Line balancing Table (Table-4).

Potential performance (G): It represents operator's efficiency level in the past on the listed operations. Select operator efficiency data against the selected operation from the skill matrix (Table-1).

Hourly potential output (H): Number of pieces expected from the operator based on his/her past performance.
   H= (Standard output X operator's efficiency%)

Time Available (I): Time available to an operator in hours at the time of starting of a task. If an operator spends 0.5 hours in an operation, he will have 0.5 hours for another operation.

Over and under production (J): It is the variation of potential output against target output. Negative (-) sign represents underproduction and no sign represent overproduction. J = (Hourly potential output – Hourly target production)

Time needed to make target production (K): Operators time needed to produce target output. K = (Hourly target production/ Hourly potential output)

Spare Time (L): Excess time after producing target quantity. L = (Time available – Time need to make target production)

Step 5: In this step, you will learn how to select an operator from skill matrix and allocate them for the operation one by one. All calculations are done for one hour.

In the Table -4, for the 1st operation (Mark Neck trim and run stitch) operator Radha has been selected as her efficiency is highest (81%) among 3 operators can perform this task. Based on her experience (skill inventory) she can produce 88 pieces per hour. So, she can meet the hourly production target and her spare time is 0.04 hour in an hour. With this much spare time we don't give her a second operation.

For the second operation (Join shoulders), operator Seema has been selected. Her potential efficiency to the job is 93% and she can make 124 pieces per hour. We need only 85 pieces to maintain continuous feeding flow in the line. To make 85 pieces Seema needs to work for only 0.69 hour. Balance 0.31 hour can be allocated for another operation that can be performed in 4TOL machine. Her excess capacity will be used later where it is needed.

Table-4: Filled format

	Operation	Output needed per hour	SMV	Standard output (@ 100%)	M/c	Operator	Potential performance (Eff %)	Hourly output	Time available	Over/under	Time needed	spare time
1	Mark Neck trim & Run stitch	85	0.55	109.1	SNLS	Radha	81	88	1	3	0.96	0.04
2	Join shoulders	85	0.45	133.3	4TOL	Seema	93	124	1	39	0.69	0.31
3	Insert neck trim	85	0.45	133.3	SNLS	Geeta	71	95	1	10	0.90	0.10
4	Serge margin	85	0.31	193.5	4TOL	Anil	80	155	1	70	0.55	0.45
5	T/S on neck trim	85	0.34	176.5	3TFL	Nirmal	75	132	1	47	0.64	0.36
6	Bind shoulder to shoulder	85	0.54	111.1	DNCS	Leela	86	96	1	11	0.89	0.11
7	Attach Sleeves	85	0.78	76.9	4TOL	Anil	77	59	0.45	-58	1.44	0.98
		58	0.78	76.9	4TOL	Sunita	80	62	1	3	0.95	0.05
8	Sew side seam with labels	85	0.84	71.4	4TOL	Roma	92	66	1	-19	1.29	0.29
		19	0.84	71.4	4TOL	Seema	90	64	0.31	1	0.30	0.01
9	Sew sleeve hem	85	0.68	88.2	3TFL	Kamal	82	72	1	-13	1.17	0.17
		13	0.68	88.2	3TFL	Nirmal	72	64	0.36	10	0.20	0.16
10	Sew bottom hem	85	0.56	107.1	3TFL	Sivani	85	91	1	6	0.93	0.07

Similar ways calculate and fill details for 3rd, 4th 5th and 6th operations. Anil who is allocated for operation "serge margin" has spare time of 0.45 hours and Nirmal who has been allocated for operation "Top stitch on neck seam" has spare time of 0.36 hours.

For the 7th operation 'Attach Sleeve' we will utilize Anil's spare time (0.45 hour) as he is working on similar type of machine (4TOL). At 77% efficiency, Anil can produce 59 pieces if he works for one hour (out of 85 pieces). In his spare time (0.45 hour), he can only produce 27 pieces. Balance quantity 58 pieces (85 – 27) need to be made by another operator. Here, Sunita (4TOL operator) is selected to make balance quantity. In an hour she can make 62 pieces (short fall is 58 pieces). Thus, Anil and Sunita's combined output meets hourly target.

Roma, who is given operation "Sew side seam with labels", can make 66 pieces in an hour. Balance 19 pieces will to be given to another operator. Seema, who is doing an operation 'Join shoulder", has spare time of 0.31 hours. As she is working similar machine type (4TOL), Roma's work is shared with Seema. At 90% efficiency Seema can make 64 pieces and in her spare time (0.31 hour) she can make 20 pieces. The combined output of Roma and Seema meets the hourly production target.

Similar ways, 9th operation "Sleeve hem" allocated to Komal and his excess work has been shared with Nirmal. As Nirmal has spare time of 0.36 hours.

The last operation has been allocated to Sivani, who can produce 91 pieces in one hour.

In this way, we can maximize utilization of operator's excess capacity and balance a line better in a better way. Instead of allocating 13 machines and 13 operators, the target quantity is achieved by 10 operators. By using this method, with balancing a line we can increase labor productivity and labor utilization percentage.

I hope you have understood the above explanation on the line balancing using skill history. This is an initial line balancing. You can use this method for initial line balancing in your factory.

How to Calculate Stitching Machine SPI?

The abbreviation of SPI is *Stitches per inch*. This means the number of stitches sewn (to be sewn) per inch of seam length. In the Metric System, it is expressed as Stitch per centimeter (SPC). In this article, I will show you how to find SPI of a stitching machine or SPI in a seam.

You may know that SPI count in a seam is a quality parameter and following the standard SPI is essential to maintain specification in a garment seam. In many garment factories, as a SOP, sewing operators check sewing machine SPI before they start sewing any garment in the morning.

SPI calculation process:

For this test, you need some waste fabrics. Then select a sewing machine for which you will be measuring SPI. Steps are -

Step-1: Take one fabric swatch of 12 x 2 inches.

Step-2: Sew the fabric length wise in a single burst with current SPI setting. Use contrast thread for bobbin and needle threads.

Step-3: Take out the stitched fabric and lay it on a flat table. Remove all creases if present on the seam line by hand.

Step-4: Take one measuring tape, place it on the stitch line on the fabric sample and mark lines 2 inches apart, as shown in the following figure.

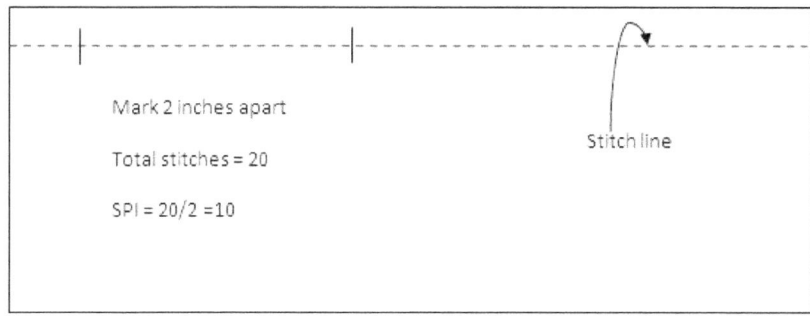

Figure: Calculating Stitch per Inch

Step-5: Count number of total stitches in between those two lines. Divide total number of stitches by 2. The calculated result is the SPI of that machine or seam. For calculating stitch per centimeter (SPC), measurement is taken in centimeters.

How to check sewing machine SPI

By following the above method, you can check sewing machine SPI. If your SPI requirement is less or more than the current machine SPI, rotate SPI regulator accordingly (for basic machines).

After SPI setting, check SPI again in the similar method. When you get correct SPI on the sample then go ahead for production.

Example: In the above figure, the total count of the stitches is 20 (in 2 inches gap). So calculated SPI will be 10.

How to Calculate an Operator Efficiency?

In apparel manufacturing, skills and expertise of a sewing operator are presented in the "Efficiency" term. An operator with a higher efficiency produces more garments compared to an operator with lower efficiency in a shift. When operators work with higher efficiency, the manufacturing cost of the factory goes down.

Secondly, factory capacity is estimated according to operator efficiency or line efficiency. Hence, efficiency is one of the mostly used performances measuring tools. In this article, I will show you how to calculate operator efficiency.

To calculate operator efficiency, you need standard minutes (SAM) of the garment and SAM of individual operations that your operator is making. Here, we will calculate individual operator efficiency.

Efficiency calculation formula: To calculate operator efficiency use the following formula.

> Efficiency (%) = (Total minute produced by an operator/Total minute attended) x 100

Total minutes produced = (Total pieces made by an operator x SAM of the operation) [in minutes]

Total minutes attended = (Total hours worked on the machine x 60) [minutes]

Example: An operator was doing an operation of SAM 0.50 minutes. In an 8 hours shift day he produces 400 pieces. So, according to the efficiency calculating formula, that operator's efficiency would be
= ((400 x 0.50) / (8 x 60)) x 100%
= (200/480) x 100%
= 41.67%

In case an operator does two different operations in a day, how do you calculate her efficiency?

Let us say, Mamta, a sewing operator, is doing two operations. In operation-1, she produced 400 garments and in operation-2, produced quantity is 380 pieces. SAM of operation-1 is 0.3 minutes and SAM of operation-2 is 0.6 minutes. What will be her efficiency percentage on that day?

Total produced minutes = (Operation-1 quantity x SAM + Operation-2 x SAM)
= (400 x 0.3 + 380 x 0.6) Minutes
= (120+228) = 348 minutes
Total minutes attended = 480 minutes
Mamta's efficiency = (348/480) x 100 = 72.5%

How to Calculate Efficiency of a Production Line?

To know the performance of a garment factory, and for the comparative study of the sewing line performance, we measure efficiency of the production lines. Like individual operator efficiency, measuring the **efficiency of a production line** is important for a factory. Daily line efficiency shows the performance level of a production line. It shows the effort done by employees and their achievement on daily basis.

The efficiency of a line for a given style can be different on each day of production. In this article, I will show you the information required for calculating line efficiency and efficiency calculation method.

Production Line Efficiency Calculation - Information required

To calculate the efficiency of a line for a day, you need the following data (information) from the line supervisor or line recorder.

1. **Number of operators** – total number of operators worked in the line in a day.
2. **Working hours** (Regular and overtime hours) – how many hours each of the operators worked or how many hours the line run in a day.
3. **Production in pieces** – Total garments (pieces) produced or total line output at the end of the day.
4. **Garment SAM** – the standard minute of the style (garment). The standard minute for the style is called SAM (standard allowed minute).

Once you have above data, calculate the followings.

- **Total minutes produced by the line:** To get total produced minutes, multiply production quantity by garment SAM
- **Total minutes attended by all operators in the line:** Multiply the number of operators by daily working hours and convert total hours into total minutes (multiplying by 60).

Now, calculate line efficiency using the following formula:

> Line efficiency (in percentage) = (Total minutes produced by the line x 100)/ (Total minutes attended by all operators)

An Example - Calculating Production Line Efficiency

Let's calculate the line efficiency of one production line. 48 operators worked in a line for 8 hours a day. They produced 160 garments and SAM of the garment is 44.25 minutes.

Here,
Total minutes produced = (160 x 44.25) = 7080 minutes
Total minutes attended = (48 operator x 8 hours x 60) = 23040 minutes
Efficiency% = (7080/23040) x 100 = 30.729%

For more examples, refer to the following table. Data calculation formula has been given on the header row of the table.

No. of Operators (A)	Working hours in a day (B)	Line output (production) (C)	Garment SAM (D)	Total minutes attended (E=A*B*60)	Total Minute produced (F=C*D)	Line Efficiency (%) (F/E*100)
48	8	160	44.25	23040	7080	30.73
48	11	240	44.25	31680	10620	33.52
34	8	300	25	16320	7500	45.96
35	11	400	25	23100	10000	43.29
35	11	329	25	23100	8225	35.61
34	8	230	25	16320	5750	35.23
34	8	200	35	16320	7000	42.89
35	11	311	35	23100	10885	47.12
34	11	340	35	22440	11900	53.03

You can also prepare a similar table in a spreadsheet for calculating line efficiency percentage. When you use spreadsheet and set the formula in the cells for calculating minutes produced, total minutes worked and efficiency percentage, you do not need to calculate it every time. You just need to fill up first 4 information – number of operations, daily working hours, line output and garment SAM.

The line efficiency calculation method shown in this article, is calculated based on the line output (produced garments). The calculated efficiency we got here is the overall line efficiency. I have shown other ways of calculating line efficiency in the next articles.

How to Show Line Efficiency when there is no Loading to a Line?

In the garment production, it has been observed that sometimes a line is filled with operators, but they were not given work to do.

It happens during the line setting or if factory has not enough cutting to load.

In such situation operators do not produce any standard minutes though they are present to the line.

Normally, factories calculate line efficiency from line output pieces. Sometimes most of operators work on the styles but as feeding get delayed, no production come out from the line on the first day of loading.

Factories calculate line efficiency in two forms - On-standard efficiency and Overall efficiency. When there is no input (loading) to the line, sewing operators will sit idle and will not produce any garment. This idle time is considered as non-productive time.

The formula used for line efficiency calculation, as explained in the previous article.

On-standard efficiency% = (Total Minutes produced / Total On-standard work hours) x 100

All hours Efficiency% = (Total minutes produced (on-standard and off-standard) /Total work hours (on-standard+ off-standard)) x 100

For the accurate efficiency reporting, record data such as on-standard work hours, non-productive time (off-standard hours), total minutes produced both in on-standard work and off-standard work.

If your line is produced nothing in whole day, then obviously line efficiency will be zero. Here you have the reason that all available hours there was no loading to the line.

If few operators work on some operations but there is no output from the, still you can show line efficiency. Calculate total minutes produced by those operators based on operation SAM. Then use above formula to calculate your line efficiency.

Let's say, in a line of 20 operators, the first 10 operators got work and they completed few bundles (produced garments on the specific operations). For making it simple, assume that first operator completed 10 bundles, 2nd operator completed 9 bundles and 3rd operator completed 8 bundles and this way let us say the 10th operator completed one bundle.

SAM of all operations is 1 minute, and each bundle contains 10 garments. Per bundle SAM is equal to 10 minutes.

Therefore, these first 10 operators combined produced minutes will be
= (10x10 +9x10 + 8x10 + 7x10+ 6x10 + 5x10+ 4x10 +3x10+ 2x10+ 1x10)
= (100+90+80+70+60+50+40+30+20+10)
=550 minutes

Total minutes attended by 20 operators
= (480 minutes x 20) = 9600 minutes
Therefore, line efficiency = (550/9600) x 100 = 5.729%

I hope now you can calculate efficiency of a production when there is no output from the line, but inside the line operators stitched garments for the new style.

Note: In case you do not capture individual operator's production, and you practice output-based efficiency, it would not be possible to calculate line efficiency when there is no line output for the style.

How to Calculate Overall Line Efficiency?

Out of different KPIs for efficiency measure, the **overall line efficiency** (simply Line Efficiency) is majorly measured by the garment factories. I will show you how to calculate overall line efficiency in a garment unit. Let me first explain what the overall line efficiency is.

Overall Efficiency

When operators are on the shop floor, sometimes they will work on the standard jobs, sometimes they will work on off-standard jobs and sometimes they may not be doing any job due to unavailability of work. In a day, each operator is given 8 standard hours for the work. The 8 hours shift time can be separated as standard work hours, off-standard work hours and lost-time hours. When the whole shift hours are used in the calculation, irrespective of how much time (hours) operators do an off-standard job and remain idle, it gives overall efficiency. When you calculate overall line efficiency, you don't need to deduct any hours.

According to the total hours, operators spend on different type of activities, their efficiency can be calculated as on-standard efficiency, off-standard efficiency.

How to Calculate Overall Line Efficiency

To calculate daily overall line efficiency, you need to measure how much hours (minutes) are produced by employees working on the line and how much time workers spent on the line for producing the output. The efficiency calculation will be done based on the line output data. Then use the below formula to calculate the line efficiency in percentage.

Let' see what all data you need to calculate overall efficiency.
- Line output - number pieces produced by the line. It may be of one style or more than one style. In the first example, I am considering line makes only one style.
- Garment SAM - Standard minutes of a garment (including machine SAM and manual SAM).
- Manpower - total manpower worked in the line including operators and helpers
- Shift hours - Total hours factory works in a day. (Normal shift is 8 hours)

You need to collect the above 4 information for the calculation.

Overall line efficiency calculation formula
The formula for calculating overall line efficiency

$$\text{Line Efficiency (\%)} = \frac{\text{Line output} \times \text{Garment SAM}}{\text{Manpower} \times \text{Shift hours} \times 60} \times 100$$

Example: Assuming that, Line-1 is making style-ABC, produces 410 pieces in 8 hours shift day. Standard minute of the style-ABC is 20 minutes and Total no. of manpower (operators + helpers) is 30.

Total minute produced= (410 x 20) =8200 Minutes
Total minutes worked = (30 x 8 x 60) =14400 Minutes (60 is multiplied to convert hours into minutes)
Line Efficiency %= (8200/14400) x 100 = 57%. Hence, the overall efficiency% of this line is 57%

You can use the following template for calculating overall line efficiency.

Line Efficiency Calculation Template

Line#	Style	SAM	Line Output	Manpower	Shift hours	Total Minutes Produced (E*F)	Total Available minutes (G*H*60)	Efficiency%
1	ABC	20	410	30	8	8200	14400	56.94
2	EFG	30	500	40	8	15000	19200	78.13
3	XYZ	15	600	18	9	9000	9720	92.59
						0	0	0.00

How to calculate overall line efficiency when a line produces more than one style in a day?

In such case, calculate total minute produced combining two styles data. Total minutes attended will remain the same.

Let's say line#1 makes two styles Style-1 and Style-2 of SAM 20 minutes and 25 minutes respectively. The production qty of Style-1 is 300 pieces a Style-2 is 200 pieces. 40 manpower work full shift of 8 hours.

Total produced minutes by line considering both styles
= (300 x 20) + (200 x 25) minutes = (6000+5000) minutes = 11000 minutes
Total minutes worked by the line = 40*480 = 19200 minutes
Line Efficiency% = (11000/19200) x 100 = 57.29%
I hope now you can calculate overall efficiency of a line.

Overall Efficiency Vs On-Standard Efficiency

You may know the definition of Efficiency. You may know how to calculate the efficiency of a production line or an individual sewing operator. But do you know -
- What is On-standard efficiency and overall efficiency?
- How to calculate on-standard efficiency and overall efficiency?
- What are the differences between these two efficiency terms?

These are the new terms to many industrial engineers and production people. If these terms are also new to you, read this post.

Workers' performance and production line performance is measured in efficiency. Efficiency is calculated using the following formula.

> Efficiency (%) = (Total minutes produced / Total minutes attended at work) x 100

When are the On-Standard Efficiency and Overall Efficiency measured?

In a normal workday, operators spend their total attended time in following activities.
- doing standard work
- doing off-standard work (doing a task which is not familiar to the operator / other than a regular job)
- doing nothing (lost time like power failure, machine breakdown, no feeding/no work)

When an operator works on a bundle (garment pieces) they produce standard minutes whether she works on-standard operation or an off-standard operation. But when an operator does nothing but sitting idle due to some reasons is lost time.

If an operator does not get work, that is not her fault. In such cases, if you measure her performance considering all the attended hours, that will reflect a wrong performance. For this reason, on-standard efficiency is measured where only on-standard work hours are considered for calculating efficiency.

Measuring on-standard efficiency is good for operator skill analysis but when it comes for production at the end of the day, or line efficiency, and incentive calculation for the line and individual operator, overall efficiency is used for eligibility level.

How to calculate on-standard efficiency, off-standard efficiency and overall Efficiency?

To measure on-standard efficiency, off-standard efficiency and overall Efficiency we need to collect the following data

(A) Total hours worked on the standard work
(B) Standard minutes produced while working on standard work
(C) Total hours worked on off-standard work
(D) Standard minutes produced while working on the off-standard work
(E) Total lost time in hours
(F) During lost time, practically, no garment will be produced, so no produced minute for lost time hours.

The standard produced minute is calculated as (operation SAM x Number of pieces stitched).

On-standard Efficiency (%) = (Total on-standard minute produced)/ (Total on-standard hours worked x 60) x100 = (B x 100)/60 x A)

Off-standard Efficiency (%) = (Total off-standard minute produced)/ (Total off-standard hours worked x 60) x100 = (D x 100)/60 x C)

Overall Efficiency (%) = (B+D) x 100/ (60 x (A+C+E))

If there is no lost-time and no off-standard work hours employee's on-standard efficiency and overall efficiency will be the same.
If there is lost time, overall efficiency will be less than the on-standard efficiency.

Efficiency Calculation (An example)

An employee produced 300 units of SAM 1 minutes in 8 hours. Out of 8 working hours, 1 hour is measured as lost time due to machine breakdown and no feeding of cuttings. Calculate on-standard efficiency and overall efficiency of this employee.

Total shift hours (8 hours) is broken down as follows

(A) Total hours worked on standard work = 7 hours (total shift hours – lost time)
(B) Standard minutes produced while working on standard work = 300 x 1 minute = 300 minutes = 5 hours
(C) Total hours worked on off-standard work = 0 (as no off-standard work performed)
(D) Standard minutes produced while working on the off-standard work = 0 (as no off-standard minute production)
(E) Total lost time in hours = 1 hour

On-standard Efficiency = (5 x 100/7) = 71.43%
Overall Efficiency = (5 x 100)/ (7+1) = (500/8) % = 62.5%

You may have observed that in somedays the line efficiency is less. Some of the reasons for efficiency loss in a plant are unmeasured lost-time, off-standard work done by sewing operators. When on-standard efficiency is measured, you can get an accurate operator performance (efficiency data). For preparing the operator skill matrix factory should consider their on-standard efficiency instead of considering overall efficiency.

The challenges in measuring on-standard efficiency are not having the right tool for capturing lost time data. With a print format, you can capture lost-time data but that would be limited, not the true lost-time hours.

How to Get Maximum Efficiency in Shorter Run Orders?

What is a short run order?

An order of small order quantity that runs on the production line for 3-5 days. The order quantity for a short-run order may be different depending on the style design. Garment factories set their own criteria for considering the order short-run or a long-run order. In the short run order, normally stitching of whole quantity completed before line reached to its pick performance level.

Why you get low efficiency for short run order?

You might have observed that for short-run orders when the line reaches its pick performance order is completed. When a factory compares the line performance of a short-run against a long-run order, it gets a less efficiency.

Normally, on the first 1-3 days of production loading (depending on the style or production system), a line builds its performance day by day (learning curve) and line works at optimum efficiency level after the learning curve. If the line gets a long run order (that normally runs more than 15 days), the average efficiency of the order comes higher as after learning most of the days, the line works on the higher level.

On the other hand, when an order gets finished by 4th to 5th day of loading, the average line efficiency remains low due to initial days' learning curve. When a line gets 4-5 short run orders in months, they lost the productive time in production line set-ups (style changeover). This results in lower line performance in that month.

How to maximize line efficiency for short run order

Here, I have shared few tips that can be useful in achieving maximum efficiency in shorter run orders. These tips are applicable for long-run orders too to achieve a better performance.
- By loading a new style back to back. This means when the previous style getting completed from the one operator, load the new style without waiting for completing style from all operations.

- Prepare the line layout and process workflow on paper in advance and follow that one during line set-up.
- All you need is, be better prepared. Keep continuous feeding to the line. Before loading the cuttings in the line, keep everything ready- all approvals related to the style, trim, and complete cutting of the order or at least two day's cutting WIP should be there.
- Train employees (sewing operators) separately, on how to do the operation they are going to assign to the style, how to handle material.
- Supervisors need to be prepared and need to complete line set-up as quickly as possible (no waiting of operators should be there), including machine changing, an instruction to operators, quality specification, and approval etc.
- Reduce bundle size. Single piece flow would be better for small runs.
- If possible, use a shorter sewing line for small qty orders. This change will also make difference in achieving line's overall efficiency.

Importance of Accurate Data in Reporting

Most of the shop floor performance matrices measured by industrial engineers are based on the garment SAM. Whether it is the line efficiency, individual employee efficiency, estimated garment cost, daily production target, or production scheduling, SAM is the primary component for making these reports.

Now the question is, how the SAM is estimated in your factory?

If you don't use a PMTS software for estimating SAM of garment operations, your estimated garment SAM may be inaccurate. Garment SAM is either less or greater than the actual SAM. In absence of PMTS software, IEs calculate SAM based on their experience.

What will happen if you calculate your performance matrices with the inaccurate garment SAM?

1. Inaccuracy in garment costing. Cost of manufacturing is calculated by multiplying SAM with an hourly rate of the labor. So, a loose SAM will give higher manufacturing cost and very tight SAM gives low manufacturing cost. Both cases are risky for the business. As you may lose your customer by quoting a high price for your products. On the other hand, you will not earn the profit in making the order if your product SAM is very tight.

2. Inaccurate performance data. When you calculate line performance and operator performance using inaccurate SAM, you will get inaccurate performance data. Repeatedly presenting efficiency in a wrong value will set a wrong perception about the factory performance.

3. Planning would be inaccurate. Factory capacity planning is done based on the standard time. All planning would be in a mess if standard minutes are not set correctly.

4. Incorrect incentive amount. If a company has an incentive plan based on individual employee efficiency, then the wrong SAM is the most critical data for a factory. Either an operator will not get fair incentive or factory will lose a lot of money in incentive without getting any return.

No matter whether you own a SAM calculator or a PMTS system or not, you can still make it right. Go to the floor, study the operators how they are working and record operation cycle time. Rate your operator accurately. Using cycle time data recalculate SAM and compare it with your SAM established for the operation. Correct it in your operation bulletin.

How to Calculate Machine Requirement for Sewing Operations?

We often need to estimate machine requirement for a new style that need to be loaded in a production line. Secondly, when one needs to prepare a project report for plant set-up, he needs to calculate machine requirement. When you prepare an operation bulletin, you need to calculate machine requirement for all operations in a style.

By reading this article, you will learn the method of calculating sewing machine requirement - how many machines do you need and what types of machines are needed to make garment in an assembly line. The primary information you need for calculating number of machines include–

1. Daily production requirement – it means how many pieces you want to stitch per day.
2. Number of hours in a shift – How many hours you plan to work each day (Standard shift)
3. SAM of each garment operation – Standard time (in minutes) for each operation of the garment
4. Average efficiency of the line - Efficiency data is required as production volume will depend on how efficiently workers may do their job. If you don't know the about the factory efficiency, you can consider industry average efficiency.

Read the following steps and practice it to learn machine requirement calculation method.

Step 1: Prepare an operation breakdown

Select one style for which you want to calculate machine requirement. Study the garment and identify the operations required to sew the garment. List down operations in a spreadsheet as per operation sequence. For example, an operation breakdown of a crew neck t-shirt is shown in the following table.

Step 2: Identify machine type

Observe the stitch class and seam types used in each operation and according to the stitch class and seam type, select machine type against each operation.

Step 3: Add SAM to each operation

In the operation breakdown sheet, write down operation SAM on the next column of machine type. SAM is the most essential data for calculating machine requirement. Operations with high SAM, need to be assigned more machines to produce daily target output. You can use SAM of each operation from your database. If you do not have database for standard minutes, then you need to calculate it before you can estimate machine requirement.

Step 4: Calculate theoretical machine requirement

Set your production target for the day (8 hours shift). Let's say, daily production requirement is 800 pieces in an operation (Sleeve attach) and SAM of this operation is 0.78 minute. Line efficiency is 80%. Shift time 8 hours (480 minutes).

This means from each operation minimum 800 garments need to be produced. Calculate machine requirement using the following formula.

Formula: Machine requirement calculation

> Machine requirement = (Daily Production Requirement x Operation SAM)/ (Shift Hours x 60 x Line Efficiency%)

Calculated machine requirement for this operation would be
= (800 x 0.78)/ (480 x 80%)
= (624/384) =1.63

It is called as calculated machine number. The formula gives you fraction of machine but in real you cannot use fraction of machine in a workstation.

Step 5: Physical machine requirement

Now round off the calculated machine number. In the above example, calculated machine requirement is 1.63. But you cannot assign 1.63 machine to that operation. Either you need to assign 1 machine or 2 machines. With one machine, you cannot achieve the daily required production. So, you need 2 machines to produce 800 garments.

For examples, I have shown an operation bulletin and operation-wise machine requirement is shown for making 800 t-shirts daily. To maximize machine utilization and assigning minimum number of sewing machine to production line, you can club operations those use similar machine class. To make target quantity in operation "Serge margin", 0.65 machine is required and for the operation "sew side seam with label" required 1.63 machines. So, you can use first 4TOL machine to do the second job.

This way you can reduce the machine number and increase machine utilization.

Table-1: Calculating sewing machine requirement (Crew neck t-shirt)

Production target/day (8 hours) =800 pieces		Shift hours = 8 hours (480 minutes)		Plan on Efficiency=80%	
Seq No.	Description	Machine Type	SMV	No. of Calculated M/c	Round-off Machine No.
1	Make Neck Rib & Run stitch	SNLS	0.55	1.15	1
2	Join shoulders	4TOL	0.45	0.94	1
3	Insert Neck Rib	SNLS	0.45	0.94	1
4	Serge Margin	4TOL	0.31	0.65	1
5	Top stitch on Neck rib	3TFL	0.34	0.71	1
6	Attach Sleeve	4TOL	0.78	1.63	2
7	sew side seam with labels	4TOL	0.84	1.75	2
8	Hem sleeves	3TFL	0.68	1.42	1
9	Bottom Hem	3TFL	0.56	1.17	1
	Total		4.96	10.33	11

Machine requirement Summary:

- Single Needle Lock Stitch (SNLS): 2 No.
- Four thread Overlock machine (4TOL): 6 No.
- Three thread Flatlock machine (3TFL): 3 No.

How to Calculate Cost of Manufacturing Apparel Products?

Estimating correct manufacturing cost is essential for a successful business. You might be preparing a garment cost sheet and you need to find manufacturing cost per garment. But you do not know exactly how to do it and you may not have full confidence calculating the same. Do not worry.

The basic method of determining the cost of manufacturing (CM) of apparel products will be discussed here.

What is the cost of manufacturing?

Cost of manufacturing is defined as the cost incurred by the factory to run the factory for making garments. Factory running cost includes direct labor costs and manufacturing overhead. Here manufacturing overheads are indirect workers' wages, staff salary, power and fuel cost, repair and maintenance cost, factory rent, administration cost etc. This cost is also known as an operating cost. Cost of Manufacturing is determined for per unit garment.

There are two ways to find the Cost of Manufacturing (CM) for a style/order.
1. Based on Standard Time (SAM) of the product
2. Based on Daily Production Average

1. CM calculation based on Standard Time (SAM)

To get better accuracy in cost estimation, one should prefer this procedure. Many small-size companies do not have such resources to measure product SAM and data for the following parameters. Following parameters are essential for cost calculation using this method.

- **Product SAM:** Standard time of the garment. The standard time of a garment is measured by using Time Study method or by using synthetic data (PMTS data).
- **Target Efficiency:** Target efficiency percentage is at what percent you are expecting running a specific product and order quantity.

- **Operating Cost per day/machine**: Operating cost is the factory running cost. Operating costs are all cost incurred to run the business other than material cost. Calculate monthly operating cost and then calculate the daily operating cost. Calculate per machine operating cost.

Formula:

> Cost of Manufacturing = ((Operating cost per day per machine)/ (Target Efficiency% x Working hours x 60)) x SAM

In the following table, an example is shown for calculating manufacturing cost using SAM and daily production data.

Table-1: Cost of Manufacturing base on garment SAM

	Particulars	UOM	Value
1	Garment SAM	Minute	21
2	Target Efficiency	%	60%
3	Actual time taken	Minute	35 (21/60%)
4	Operating cost per day per machine (example)	₹	1022
5	Normal working hours/Day	Hours	8
	CM cost per piece	(1022/480) x 35	74.52

In the above example, garment SAM is 21 minutes, target efficiency 60%. So, the actual time would be 35 minutes to make one garment. The factory works 8 hours a day and operating cost per day per machine is ₹ 1022.

Therefore, cost of manufacturing one garment is ₹ 74.52

2. CM calculation based on daily production data

This method is widely used by garment manufacturing factories. Cost of Manufacturing calculation is done based on historical production data. This is an easier method compared to the above one.

Information required to find Cost of Manufacturing

- **Daily production:** Find the average daily production of a style (garment) based on earlier (historical) production data. Calculate the daily average production of the factory.

- **Manpower involved in production:** Number of sewing machines or sewing operators are utilized to produce the above quantity.
- **Operating cost per day/machine:** As explained above.

Formula,

> Cost of Manufacturing = (Operating cost per Day / Total garments to be produced per day)

See the example method -2 in the Table-2. In the example, the daily production is 550 pieces. 40 operators worked to produce these pieces. Operating cost per machine is ₹ 1022 per day.

Table-2: Cost of manufacturing based on daily production data

	Particulars	UOM	Value
1	Daily production	Pieces	550
2	No. of operators	No.	40
3	Operating cost per day per machine	₹	1022
	CM cost per piece	(1022 x 40)	₹74.33

So, the cost of manufacturing is ₹74.33

P.S. Data used in the above examples are hypothetical.

How to Calculate Number of Sewing Lines Needed for an Order?

Question: *How can I determine number of lines required from the value of order qty, SMV, Working hours and days in hand to shipment?*

In a factory, production planner and IE plan for resource requirement in terms of manpower requirement, machine requirement and material requirement for the new orders. They also plan number of production lines to be used for an order to complete the order on time.

Here, I will explain how to calculate number of sewing lines for an order when you have information of order quantity, product SMV, and number of days in hand for production.

To determine the number of sewing lines that are needed to complete an order on time, following information is required.

- Order Quantity
- Garment SMV
- Working Hours per day
- Days in Hand for production
- Line Efficiency and
- No. of sewing machine installed in each line.

To find out number of sewing line follow the below steps.

Step -1: Calculate daily production requirement.

When you are given an order and you know how many days are available to complete stitching, you can calculate daily production requirement. Use the following formula to calculate daily production requirement.

> Daily production requirement = (Order Quantity / Number of days in Hand)

It is considered that days in hand are only for production, not up to finishing and packing. We will get equal production requirement for each day using this formula. This is okay for the purpose of calculation. In practice, there will be learning curve when the style will be loaded, and actual production will be different in day-1, day-2, and day-3.

Another note - when you calculate number of days in hand, only count working days excluding weekly off and holidays.

Step-2: Calculate production capacity of each line.
Use the following formula for calculating line capacity in terms of pieces.

> Line capacity = (No. of machines x Working hours x 60 x line efficiency%) / Garment SAM

When you calculate daily line capacity, consider the learning curve. You need to follow learning curve efficiency for calculating daily production capacity from day-1, day-2, day-3, etc. Learning curve must be considered here. From the historical data, you need to find learning curve efficiency of your factory.

Step-3: Calculate number of lines required
Divide daily production requirement by average line capacity to calculate number of lines to be set to complete production before production completion date.

> Number of Lines= (Daily production requirement/ Line capacity)

Case: Assume that you have received an order of 50,000 pieces of a style with 15 minutes SAM. Your factory works for 8 hours day and you have 20 days production time in hand. You have 30 machines per line which performs at 50% efficiency. In this calculation, I am not using learning curve. That means from day-1, line will work at 50% efficiency (for calculation purpose only). In another article, I have shown daily capacity calculation using the learning curve efficiency and calculating number of days required for completing an order.

Step-1: Daily production requirement = (50,000 / 20) = 2500 pieces per day.
Step-2: Line capacity = (30 x 8 x 60 x 50%)/15 = 480 pieces per day
Step-3: No. of lines = 2500/480 pieces =5 lines (5.2)

I hope now you can calculate the number of production lines you need for an order to complete the production on time.

How to Reduce Line Setting Time?

In a garment factory, engineers and production managers always look for a way to improve the factory's labor productivity. But many times, they look over the things that can reduce labor productivity.

"Higher line setting time" is one of the most visible reasons that reduces factory's overall productivity. When it takes a longer time for setting a line, most of the operators sit idle. That means operators are not utilized in producing garment and operator productivity falls resulting high labor cost.

I have seen factories where 1.5 to 2 days are spent in line setting for woven tops. When line supervisor and engineers are asked why they are taking that much time to set a 40 machines line, they give dozen of reasons. Longer line set-up time cause production delay. Reasons may vary from time to time or style to style.

Why line setting takes a longer time?

Followings are some of the reasons cause longer time for line setting.

1. Factory starts loading a new style to the line once all operators get free from the previous style.
2. Frequent change in line planning.
3. Starting line setting without ensuring that all required trims are approved and sourced. Until required trims are sourced all operation cannot be started.
4. Depending on the style design, sometimes garment components are sent outside for printing or embroidery process. When there is delay in receiving garment components from outsourced process.
5. When supervisor does not fully assess the operation sequence or skill requirement for each operation.
6. Operators who were supposed to be assigned for the new style are absent on the line setting day.
7. The quality issue, a supervisor not able to give a suitable operator for the critical operation
8. Maintenance person is not able to set-up the machine quickly. Replacement of machine, setting guides and attachment takes a longer time than it should be.

9. Planning for larger bundle size. At the first day of line setting if a bigger bundle size is used then it will take a huge amount of time to reach the bundle at the last operator.

Tips to reduce line setting time

Work on the above reasons and eliminate them prior to starting the line set-up. Once you know the reason you can resolve it. In the following list, a few tips are given to reduce line setting time.

1. Research and development of the style – study the style well before setting up the line. By doing so, you will be aware of critical operations, machine requirement, skill requirement for the operations.
2. Check the production file thoroughly at the time of receiving from merchandising team – check whether trims are approved or not. If any items are not sourced yet when it is expected. Plan line setting according to availability of goods.
3. Prepare line plan with manpower requirement for specific skill categories. Ensure that operators selected for the operations are present during line setting.
4. Arrange all necessary attachments, needles, guides you need well in advance and tested in sampling or Research and development center.
5. Allot dedicated machine maintenance team and quality personal during line setting. Machine set-up and stitching quality checking can be done outside of the line.
6. Make small size of bundles (3 to 5 pieces per bundle) for the first day of loading a new style. Thus, bundle will move fast at the end of the line. Once WIP is build up, bundle size can be increased.
7. Use machine shifting device for replacing machine quickly on the floor.
8. Prepare OB sheet, line layout in advance. Communicate with line supervisor, machine maintenance team for the upcoming styles. Ask them to arrange all required machines, equipment, guides, presser foot, and attachment in advance.

By following the above guides, you can reduce line setting time.

How to Control Apparel Production Cost?

When it comes for controlling apparel manufacturing cost, there are many ways to reduce the apparel production cost that you can follow. In this article, I will show you 4 ways to reduce and control manufacturing cost.

I was in a meeting discussing about line efficiency, present cost per pieces and what the production people can achieve to lower down the making cost. When we started discussing the fact everybody was surprised. Read the following case study to know what the surprising thing was.

Case Study:

Product SAM = 23 minutes (Ladies blouse)

Hourly salary per operator = ₹40.00 approximate (Assuming operator's monthly salary ₹8000.00, and 25 days month)

Present output (on 4th day of line loading) = 300 pieces 40 operators work 8 hours in a day for producing those pieces.
So, present labor cost per piece = (Total cost of operators/number of pieces produced) = (40 x 40 x 8)/300 = **₹42.67**. It seems fine.

According to the standard minutes (SAM of the garment), if line performs at 100% efficiency then it would cost factory = (40/60) x 23 = ₹15.33 per piece [(Production cost per hour/60) x SAM].

Even if the factory hits average line efficiency to 50% operator cost per pieces become = ₹15.33 x 100/50 = **₹30.67**.

There is a big difference between current performance and what is expected.

Assume that the factory reaches at 50% average efficiency then factory can save net ₹12 (₹42.67 – ₹30.67) per piece. In an order of 15000 pieces, factory will save approximately ₹180,000 in this order. See the below chart, production cost per piece reduced when line efficiency is increased.

In the following chart, production cost per garment is shown at different line efficiency level.

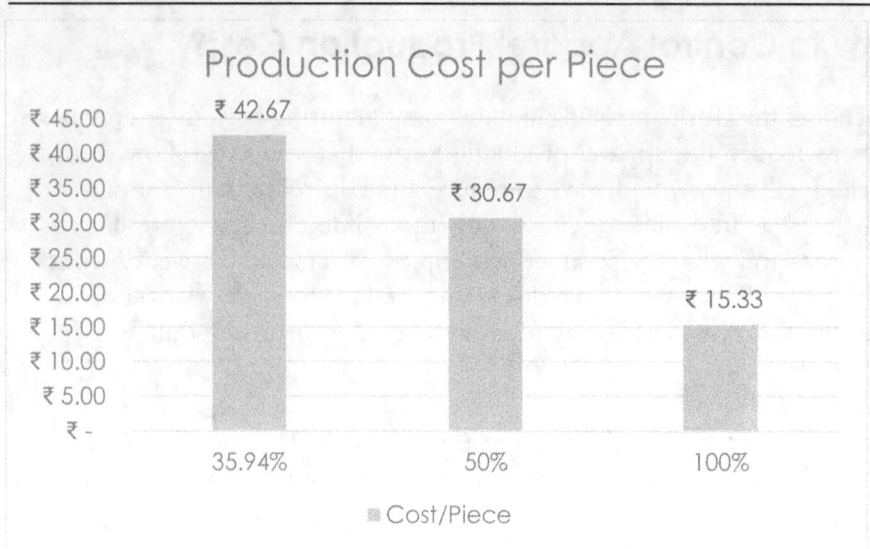

Chart – Line efficiency Vs Production cost per piece

These figures are all real and can be achieved if you know how to achieve the target efficiency. Now question is how to improve the line efficiency. Total working hours in a day are fixed. You need to produce more minutes (more pieces) to increase the efficiency within shift hours. The initial steps to start improving line performance are listed below.

1. Increase operator utilization through balancing the line and eliminating off-standard works or minimizing lost time.
2. Train operators on specific operations to improve their skills
3. Motivate employees to perform jobs efficiently
4. Stabilize line by minimizing operator absenteeism and turnover

You can work on the above things to meet your target (lowering the production cost). Otherwise, you can seek for professional help. Now-a-days you will find many freelancers who can help you saving money at your factory day by day.

Conclusion:

There are many ways to improve efficiency of a production line. Many of those are explained in the other articles in this book. You can also follow those tips. It is true that by increasing production volume using same resources, you can control production cost.

How to Calculate Sewing Floor Capacity?

When we talk about sewing floor capacity, it means number of garments a factory can produce per day for a given product. And the production capacity of a factory depends on product type, availability of sewing machines and manpower resources. You can calculate whole factory capacity considering a single product category. Or you can first calculate line-wise and product wise production capacity, then floor-wise capacity and finally calculate the factory capacity.

In this article, you will learn the method of calculating the available capacity of the sewing line. The available capacity of a line is presented in minutes or hours.

> **Sewing room Capacity per Day**= {(No. of machine x daily work hours x 60) – absenteeism %} x Efficiency %

> **Monthly Capacity** = (Daily Capacity x No. of working days in a month)

The capacity of a line depends on the average efficiency% of the line and operator absenteeism%. First, line-wise capacity is calculated and then all lines' capacity data are added together to calculate the available capacity of the floor.

We need the following information for calculating line capacity

- Number of sewing machines available in a line
- Shift hours in a day (standard shift hour -8 hours)
- Average operator absenteeism percent
- Average line efficiency%

An example: A factory floor has 5 lines. The factory works for 8 hours daily. No of total operators, line efficiency, and absenteeism percentage are as given in the following table.

Table-1: Line-wise capacity calculation (in minutes)

Line No.	No. of Operator	Minutes/Day (Daily working Hrs. X 60)	Line Efficiency	Absenteeism %	Capacity available (Minutes)
Line 1	30	480	40%	10%	5184.0
Line 2	28	480	50%	10%	6048.0
Line 3	32	480	55%	10%	7603.2
Line 4	32	480	45%	10%	6220.8
Line 5	25	480	39%	10%	4212.0
		Total Sewing floor capacity per day (in minutes)			29268.0

The available capacity of the line will vary on the factors such as

    i)   Number of operators working in the line,
    ii)   Line's existing efficiency and
    iii)   Operator absenteeism percentage.

Capacity can be also represented in pieces. To capacity in pieces, divide total minutes capacity by SAM of the garment that will be made on the floor.

Assume that a factory produces basic full sleeve shirt of SAM 20 and it has 29268 minutes available capacity. Then shirt production capacity of the floor will be 1463 pieces per day (29268/20).

> Production Capacity in Pieces = (Available floor capacity in minutes/Garment SAM)

How to Calculate Production Capacity of a Factory?

The capacity of a factory is primarily expressed in terms of total machines a factory has. Another way of presenting capacity of a factory is the number of pieces it produces on daily for the specific products. In general, total numbers of machines in a factory mostly remains the same for a period. When a factory makes different types of the product during the season, machine requirement may change and the daily average production capacity in each style may vary.

So, to be specific during booking new orders, a planner should know exactly how much capacity they need to produce the whole order within the given lead time.

A factory's capacity is presented in total minutes or hours or in pieces (production per day). In the previous article, you read how to calculate sewing line capacity. The method used to calculate capacity of a factory is explained in the following. To calculate daily production capacity (in pieces) you need the following information.

1. Factory capacity in hours
2. Product SAM
3. Line efficiency (Average)

Step-1: Calculate factory capacity (in hours)

Check how many machines factory has and how many hours factory runs in a day. An example: assume that total number of machines 200, Shift hours per day is 10 hours.
Therefore, total factory capacity (in hours) = 200x10 hours = 2000 hours.

Step-2: Calculate of product SAM (Standard Minutes)

Make a list of the product category that you manufacture and get standard minutes (SAM) of all products you make from work-study engineers. If you do not have product SAM, then calculate the SAM. Let's say you are making a shirt item and its SAM is 25 minutes.

Step-3: Calculate average efficiency of the factory

If you are a production planner, collect these data from the industrial engineer. Or calculate average line efficiency from the historical data. Suppose average line efficiency is 50%. In another article, I have explained how to calculate efficiency of a production line.

Step-4: Calculate production capacity (in pieces)

Once you have the required information, use the following formula to calculate production capacity in pieces.

> Production capacity (in pieces) = (Available capacity in hours x 60/product SAM) x Line Efficiency

Example: A factory has 8 sewing lines and each line has 25 machines. Total 200 machines and the working shift is 10 hours per day. Total factory capacity per day is 2000 hours (200 machines x 10 hours). The factory is producing only one style (Shirt) of SAM 25 minutes and used all 200 machines, daily production capacity at 50% efficiency will be
= (2000 x 60/25) x 50% Pieces
= (2000 x 60 x 50) / (25 x 100) Pieces
= (6000,000/2500) Pieces = 2400 Pieces

Table-1: Production capacity calculation template

	Number of total machines	Shift hours	Product SAM	Average line Eff%	Production capacity
	\multicolumn{5}{l	}{Case-1: Calculating production capacity for single product}
1	200	10	25	50%	2400
2	200	8	25	60%	2304

[Note: Production capacity will vary according to the line efficiency and during learning curve]

Production (capacity) planning is normally done based on sewing capacity. Having knowledge of the capacity in other processes (internal or external) is also very important. Otherwise, the planner may fail and will not be able to meet the deadline. Other departments such as cutting room capacity, finishing room capacity, washing capacity, and capacity of the value-added jobs can be calculated using the same method.

Questions related to production capacity calculation of a factory in various situations

Over the time, I received many questions on capacity calculation. I have added some selected questions here. Hope you will find these questions & answers helpful.

Q-1: Sewing line includes sewing machine as well as pressing machine. Do we need to take all the machine or only sewing machine for production capacity?

Answer: In the production capacity calculation formula, we are using product SAM to calculate the daily production capacity. Therefore, if you are adding finishing operation SAM, manual work (like marking, ironing) SAM, in the total garment SAM, those workstations will be part of the calculating factory capacity working hours. But as some of these workstations varied from style to style, considering the sewing machines for calculating machine capacity in hours.

Q-2: How do we estimate this 50% efficiency? And how does this influence the SAM?

Answer: The line efficiency that I used here comes from the past data of the line performance (it is an assumption). SAM is the standard time of the style. Standard time will remain the same for a product. In the above example, at 50% efficiency, daily production target is 2400 pieces. If the line efficiency is 75%, daily production quantity will be 3600 pieces.
Garment SAM will not change whether the line work at 50% efficiency or 80% efficiency. The production capacity of a line will change if line efficiency varies.

Q-3: Given the formula above. what if you have 5 products with different SAM. Product A - 5 minutes, product B - 10 minutes, product C - 15 minutes, product D - 20 minutes, and product E - 25 minutes How can I compute the production capacity?

Answer: You can compute production capacity of a factory making product of different SAM. Here you know.
- Product SAM for all 5 different products.

- Find product wise average line efficiency.
- Calculate total capacity in hours. You already have the capacity data machine hours daily. (shown in the step#1)

Out of the total available machine hours, you need to decide how many hours you will allocate for different items. See the product wise production capacity Table-2. I am assuming that production wise machine allocation, daily shift hours, and line efficiency as mentioned in the Table-2. Then follow the formula and find product wise production capacity. While calculating production capacity, always consider the learning curve.

Table 2: Template for production capacity calculation for multiple products.

Case-2: Calculating production capacity for multiple products						
	Product A	Product B	Product C	Product D	Product E	Total
Number of machine	50	50	100	100	100	400
Product SAM	5	10	15	20	25	
Line Efficiency	70%	80%	90%	70%	80%	
Shift hours	8	8	8	8	8	
Product wise production capacity (pieces)	3360	1920	2880	1680	1536	
Total factory capacity (pieces)						11376

Q-4: It will not work if number of products are very much. Say there are 100 type of products and all products manufactured in the same machine and takes different time and there is no certainty of the requirement of the product then how someone can calculate installed capacity.

Answer: It is normal that the product types that you make will change from time to time. SAM of the same product may also change. Production capacity in pieces (dozen) will change accordingly - if the product SAM and line efficiency changes. The number of machines is fixed, and one machine can be used for making one product at a time.

In such a scenario, you do not fix/show the factory capacity in pieces. You can keep your machine capacity in hours per day, as it will remain the same.

Calculate the production capacity for the current orders / current styles only or for the upcoming styles. Prepare a separate capacity report for each product type you are making with the time frame. For an example, suppose you have 100 sewing machines and currently you are making a t-shirt and shirts only. Out of those 100 sewing machines, 20 machines are allocated for making t-shirts (5 minutes SAM) and 80 machines are allocated for shirts (20 minutes SAM). T-shirt line work at 90% efficiency and shirt line work at 70% efficiency level.

Therefore, your factory capacity would be as below. (Table -3)
- 100 sewing machines,
- 800 hours per day,
- 1728 units of T-shirts and 1344 units of shirts production capacity per day.

Table-3: Calculate product wise capacity by changing parameters

	Product A	Product B	Total
Number of machines	20	80	100
Product SAM	5	20	
Line Efficiency	90%	70%	
Shift hours	8	8	
Product wise production capacity (pieces)	1728	1344	
Total factory capacity (pieces)			3072

After a month, suppose you got an order for making Polo shirt and a trouser style. So, in 20 machines allocate the Polo shirts and other 80 machines make the trousers. Calculate factory production capacity separately as above. You can increase /change the number of machines for the products as per requirement. If you get an order of multiple items at the same time, you calculate product wise capacity in pieces according to the machine you will be allocating for each style.

Conclusion:

Production capacity calculation is a common task to be done by an industrial engineer. You may have question on how to calculate factory capacity on various situations. I have already answered few questions in this article. Go through those answers. I am sure you will get answer to your questions as well. Also read, other articles in which I have explained factory capacity calculation method.

How to Calculate Hourly Production Target?

In apparel production, we often need to calculate hourly production target for a given style for setting up target for production lines and individual operators. The hourly production target is calculated for individual operators as well as for a line, as needed. I will show you the following -
1. How to calculate hourly production target for an individual operation?
2. How to calculate hourly production target for a production line?

Hourly production target calculation is a part of an operation bulletin preparation and daily production report. In an OB production target is calculated for each operations. Following formula is used to calculate hourly target for individual operations.

> Hourly Production Target = (60/Operation SAM)

An hour is equal to 60 minutes. Divide the 60 by the operation SAM. The result of this calculation tells you the number of pieces can be produced per hour, which is the hourly production target for the selected operation.
Let's say SAM of an operation is 0.5 minute. Hourly production target would be
= (60/0.50) =120 pieces

This target is calculated at 100% efficiency. In case you want to plan hourly production target at the certain percentage, multiply the above by individual operator efficiency. When you calculate operation wise hourly target, you may not know the accurate efficiency level of the operator who will be working on the operator. In such situation consider average line efficiency data.
If the skill level (efficiency) of that operator is 80%, production target would be = (60/0.50) x80% =120 x 80% =96 pieces
These two information you need for calculating hourly production target for one employee.
- Standard time (SAM) of the operation

- Efficiency of the individual operator (or average line efficiency)

Calculation of hourly production target for a line

On the other hand, for calculating hourly production target for a production line use the following formula.

> Hourly Production Target of a line = ((60 x No. of operators working in a line)/Garment SAM) x Line efficiency %

You see to calculate hourly production target of a line you need these information –
- Garment SAM
- Average Line Efficiency
- Number of employees working on the line

If you need to calculate daily production target of a line, just multiply the hourly line target by shift hours.

Question: Calculate hourly target and daily production target when a line works at 45% & 60% efficiency.

We will calculate hourly production target and daily production target using the following case.
- Garment SMV=11.25,
- Number of Worker=25,
- Shift Hour=10.

What would be the hourly and daily production target at 45% & 60% efficiency?

First calculate hourly and daily production target when line efficiency is 45%

The given garment SMV is 11.25 minutes. First, calculate garment SMV @45% efficiency. You know, SMV will be higher than 11.25 when efficiency is less than 100%. Therefore, SMV @45%

= 11.25/45% = 25 minutes

Hourly Target is calculated using this formula

> Hourly Target= (Available minutes in an hour / Garment SMV)

In this case, 25 workers are involved in making the garment, so the total available minutes in an hour would be 25 × 60 minutes.
Therefore, the hourly production target of the line @45% efficiency will be = (25 × 60/25) = 60 pieces
The production line works 10 hours a day.
So, the daily production target will be 10 × 60 = 600 pieces

The calculation is shown in the following table.

Manpower	SMV	Line Efficiency%	SMV @ target Efficiency%	Target per hour	Shift time	Target per day
25	11.25	45	25	60	10	600
25	11.25	60	18.75	80	10	800

Now, calculate hourly and daily production target when line efficiency is 60%. Garment SMV @60% efficiency = 11.25/60% = 18.75 minutes
    Therefore, @60% efficiency, hourly target will be
    = (25×60/18.75) = 80 pieces.
Production target in 10 hours would be 800 pieces (80×10).

How to Calculate Sewing Operator Requirement?

Prior to setting up a sewing line for a new style, you need to calculate manpower requirement. Normally, estimated manpower requirement is shown on the style's operation bulletin. The operator requirement is calculated for each operation.

The number of operator requirement per operation is calculated using this formula.

> Calculated manpower = (Operation SAM x Production requirement per shift) / (Total available minutes in a shift x Target Efficiency)

Let's assume

- Daily production requirement is 500 pieces,
- Operation SAM is 1.5 minutes,
- Factory works 8 hours a day (i.e. 480 minutes) and
- Target line efficiency is 70%.

Here, calculated manpower = (500x1.5)/(480x70%) = 2.2

For your understanding, I have shown four more operations of a T-shirt. No. of calculated operators and allocated operators are shown to produce 2400 pieces per day at 75% target efficiency in 8 hours shift.

The calculated operator number often comes in decimals. But in practice, we cannot assign a fraction of an operator to any operation. So, use the nearest round number for calculating operation wise manpower requirement.

Table-1: Manpower requirement calculation for a sewing line

Operation	Machine type	SMV @ 100%	Calculated Manpower	Allotted Manpower
Shoulder join with tape	4T OL	0.170	1.1	1
Wash care label tack	SNLS	0.220	1.5	2
Neck rib joining to body	4TOL	0.500	3.3	3
Top stitch on neck seam	Flatlock	0.200	1.3	1

Grouping of operations:

Remember, the calculated manpower is not the final manpower requirement for the given style. You need to look at each operation and check whether an operator is utilized fully or not. You might find that there are some operations which do not need 8 hours of an operator to produce the target quantity. In such cases assign multiple operations to the operator. To do this you also need to check machine types required for doing the selected operation.

In a critical operation, you may need to add more than one machines (higher than the calculated machine numbers) and thus need an additional manpower.

All this decision is made by the engineer to achieve the target production. Simply following the calculated data for manpower and machine allocation may build up WIP in some operations in the line and form bottlenecks.

Note: When we calculate manpower requirement for a sewing line or for a factory, we consider the operator absenteeism factor in calculating total manpower requirement for the line. The same can be used in calculating operation wise manpower calculation in an operation bulletin.

How to Estimate Machine Requirement for a Factory?

In mass garment production, various types of sewing machines are used. Types of machine mix one should purchase depends on the product category to be made by a factory. As you know for knitted garments, you need a higher number overlock and flatlock machines but for woven products you need a higher number of lock stitch machines.

When you plan for a new factory set up for garment manufacturing, you should select machines based on your product/item categories. Otherwise, you will have a wrong machine ratio. As a result, you will fall short of some types of sewing machines and other machines will be less utilized. It is not only the selection of correct types of machines, but you should also have the correct number of machines for all type of machines you purchase. If you purchase machines only those are required initially then you can control in initial capital investment.

Here are the steps of how I select machine mix for a new factory setup.

1. **Select Product Type -** As I mentioned above, the requirement of machine types depends on product types and product styling. So, first thing is you need to decide the product. Is it Men's shirt, Blouse, Tee-shirt, Trouser, Salwar Kameez or Kurtis? To make a t-shirt, you need 3 different types of machines but to make a trouser you need 7 different types of sewing machines.
2. **Daily Production Target:** Once you finalize the product or product groups that you wish to make, decide how many pieces of garment you like to produce daily. In another way, use the average production demand from your customers. As an example, let's say you plan to produce 1200 shirts per day.
3. **Estimated Line Efficiency:** At the start-up stage you would not be knowing the line efficiency. But you need to consider an efficiency figure at what level your line may perform 6 months ahead to calculate machine requirement. For this consider industry average efficiency data. As most of the factories run at 40% efficiency (in India for woven products), use this figure for calculations. If you want to consider a higher efficiency level, you can. For a knitted t-shirt, line efficiency can be taken as 80%.
4. **Prepare Operation Bulletin:** Next step, you need to prepare an operation bulletin (OB) of the selected product. If you know it,

study the product and check what type of operations are needed to make the garment, select correct types of machines for each operation and then calculate product SAM, number of machines for each operation, machine requirement per line to get production as per your target. For multiple products, prepare OB for each product and calculate machine requirement accordingly.

5. **Calculate number of sewing lines:** Divide daily total target by per line estimated production to calculate the number of sewing lines. For example, if you plan for 400 pieces per line per day then you need to set 3 production lines to produce 1200 pieces per day at the defined line efficiency.

6. **Finally, make a matrix of machine mix:** Once you have line wise calculated number of machines for each product, prepare a table with machine list of each product. Calculate total number of machines for each type. Refer to the following Table-1 for the machine mix matrix.

Following the above steps, you can calculate machine requirement for a new plant. Here, I have shared one sample machine mix report for making trousers, shirts, t-shirts, and polo shirts. In the Table-1, 13 different types of machines are planned for 16 lines for making these 4 products.

Table-1: Machine requirement sheet

Sl. No.	Machine Types	Product Categories				Total
		Trousers (2 lines)	Shirts (4 lines)	T-shirts (8 lines)	Polo-shirts (2 lines)	
1	SNLS with trimmer	50	112	14	26	202
2	SNLS with edge cutter		12			12
3	3T Overlock	5			12	17
4	4T Overlock			32		32
5	Single needle chain stitch	4				4
6	Feed of the arm (FOA)		6			6
7	Keyhole making machine	2				2
8	Buttonholing	2	6			8
9	Button sewing		6			6
10	Bar tack machine	2			1	3
11	Snap button	1			1	2
12	Flatlock (flatbed)			20	6	26
13	Flatlock (Cylinder bed)			4		4
	Total	66	142	46	46	300

How to Calculate Helper Requirement for a Sewing Line?

Question: *How many helpers to be placed in a line of 60 machines layout & 45 machines layout? What is the method for calculating above? ...asked by an OCS reader*

There is no formula for calculating number of helpers to place to a production line. Helpers are normally hired for a line to do manual tasks for a style, to feed cuttings to a line. Like you calculate number of sewing machinist (and sewing machines), number of helpers for manual operations can be calculated based on SAM of manual operations.

Following are the guidelines that will help you to decide how many helpers you need to hire irrespective of a line's strength.

1. Go through your operation bulletin and check in how many jobs /tasks you need helpers (like marking, trimming jobs, ironing job), for some manual task you may need multiple helpers.
2. **Check feeding procedure:** To feed cutting to line do you need any helpers? If yes, do you need one person for each line, or one person can feed bundles to multiple lines? Calculate number of helpers based on your requirement.
3. **Matching of parts:** In many styles, part matching is required prior to feeding garment parts to the assembly section. This is done to eliminate shade variation and avoiding size mixing. This also helps in saving operators time who will be doing initial assembly operations.
4. **Sorting of bundle:** Do you need to sort bundles in the line?
5. **Hand work:** Does your style need manpower for handwork?
6. **Thread trimming from garment:** Do you need helpers for thread trimming in the sewing line? If yes, count thread trimmers as helper.

To estimate the total number of helper requirement for a line, follow the above guidelines. You will get an approximate number of helpers. For an example, 1 helper for feeding cuttings, trims and accessories. Plus, other helpers for manual tasks for the line.

How to Calculate Manpower Requirement of Finishing Section?

Question: *How to calculate manpower strength required in the garment finishing department?... asked by an OCS reader*

You know the processes involved in the garment finishing section. Processes include thread trimming, garment checking (initial and final checking), measurement checking, spotting (stain removing), pressing garments, folding, and touching.

To determine manpower requirement in the finishing section, follow the below steps.

1. **Define** what all works/tasks that you perform in the finishing section. Like checking, thread cutting, pressing, measurement, folding, etc.
2. **Do Time study** for each work with your current workers and do capacity study of your finishing room workers for each types of work. For example, as per your capacity study a worker can check 10 pieces/hour, a folder can fold 25 pieces/hour. Thus, a checker checks 80 pieces per day and a folding-man folds 200 pieces per day, respectively.
3. **Now find Daily Production Target** for finishing section. (Suppose finishing target 800 pieces daily).
4. **Calculate manpower requirement:** Manpower required in a finishing process = (Target quantity/capacity per worker per day).

   From the above example, you need
   Total checker= 800/80 = 10 no. and
   Total Folder man = 800/200 = 4 no.

You see you need to know daily finishing target and finishing capacity per operator in each finishing process. Calculate process-wise manpower requirement and then calculate total manpower requirement for the finishing section.

Relationship Between Machines, Working Days and Output

Question: *How to calculate no. of machines, working days & output based on order quantity? ... asked by an OCS reader.*

First, let me explain the relationship between the number of machines, total working days and output for an order.

Daily output of a production line is directly proportional to the number of sewing machines used for production. And total working days required to complete an order is just reverse (provided factory works daily fixed hours). This means if you increase the number of sewing machines, number of required days in completing an order will reduce.

To calculate these parameters, we use following formula

1. Daily projected output **(A)** = [(No. of machines in a line **x** Shift hours)/ (Garment SAH/line efficiency%)]**x** No. of lines

    [Garment SAH = Garment SAM/60]
    To calculate daily output, you need these information
    - Number of machines you will set to the line
    - Garment SAH (SAM/60) and
    - Line efficiency

    You may set-up multiple lines for making one order.

2. Total working days **(B)** = Total order quantity/Daily output **(A)** (Daily output to be calculated considering the learning curve)

3. No. of machines **(C)** = (Daily output required **(A)** x Garment SAH)/ (Efficiency x Shift hours)

    From the above three formula, now you can find relationship or calculate A, B and C from order quantity (OQ). But with garment SAM, Line efficiency and shift hours, you must have data one of the above three parameters to calculate other two parameters.

4. No. of machines **(C)** = (Order quantity **(OQ)** **x** Garment SAH)/ (Efficiency% **x** Total working days **(B)** **x** Shift hours)

5. Total working days **(B)** = Order quantity **(OQ)**/Daily output **(A)**
The same is shown in step#2

6. Daily projected output **(A)** = Order quantity **(OQ)** /Total working days **(B)**

For example, An order quantity of 12000, garment SAH 0.5, line efficiency 50% and shift hours 8.

If factory has 50 **(C)** machines in a line and the order is made in that line only, then A = 400 and B = 30

If factory produces 400 (A) pieces daily then, B=30 and C=50
If production duration is 30 days (B) then, C=50 and A=400

How to Calculate CM Produced by a Line in Dollar?

Question: *Sometimes people ask me "how much CM (in USD) is achieved today in your factory". It seems confusing to me. Would you please tell me how can I calculate CM in daily basis? And why I need to calculate every day?*

On daily basis you calculate line output. And you have the SAM for the style(s) you are making in the line. With these data you calculate how much SAM is produced by the line.

> SAM Produced by line = (Line output **x** Style SAM)

If you have this information, calculation of **CM achieved by the line in dollar** is an easy task. I will show you two different ways for calculating CM earned amount by the line. These methods can be used for calculating CM achieved by the factory as well as CM achieved by the line.

The first method:

In the costing sheet you include product CM cost, which is your garment manufacturing cost per unit. CM cost is part of FOB. (CM stands for cost of manufacturing).

Most of the time, CM cost per garment is shared with buyer (in open costing). So, you need to know CM cost per piece for the product that you are making.

Apparel buyers pay CM on total quantity shipped to them. Therefore, every piece a factory produces are virtually earning dollar equivalent to CM.

In this method, CM earned is calculated as total line output (pieces produced by the line) multiply by CM cost per garment.

If you want to calculate total CM achieved by the factory, find total production done by the factory and multiply it by CM cost.

Calculation method:

1. From the cost sheet collect garment CM (CM is fixed for an order). This data is normally available with merchandisers. In case, cost of manufacturing is in local currency, convert it into dollar using currency exchange rate.

2. Count total piece produced by a line
Now calculate CM achieved (in dollar) = (Production quantity × Garment CM in dollar)

The second method:

There is a second way of calculating CM earning (CM achieved). 'CM achieved' can be defined as total calculated earned amount by a line which derived from total minutes produced by the line multiply minute cost.

In apparel production, when you are running a line, direct cost is involved in operators' salary. And every factory has a minute cost (labor cost per minute for direct labors).

To make it simple, you are paying salary to your operators and workers. In return operators are producing garments. To compare it with how much dollar you get in the return in dollars by paying operators wages on a day, CM earning is calculated.

Calculation Method:

Here are the steps to calculate CM achieved by a line:

1. Calculate labor cost per minute of your factory based on minimum wages (in USD) of the workers at 100% performance. This factor changes when worker's salary is changed, otherwise cost (dollar) per minute remain same.

2. Calculate total standard minute (SAM) produced by the line. Again, there are two different ways to calculate Standard minute produced–
   (a) Standard minutes produced = (Line output × garment SAM)
   (b) Standard minutes produced = Sum of total minutes produced by each operator in a line.

3. Calculate total hours worked by all workers (employees who are included in the SAM)

4. Dollar earned (dollar achieved) = (Total standard minute produced × Dollar per minute)

Additional analysis for your information:

5. Dollar spent/paid = (Total minutes worked × Dollar per minute)

6. Excess paid (Make up amount) = (Dollars paid - Dollars earned)

On daily basis calculate total standard minutes produced by a line and total garments produced by the line. I hope you do this every day. So, whenever you are asked this question, just follow one of the above methods.

To make the calculation simple, follow the first method. Multiply CM cost (Dollar per garment) to the pieces produced by a line to calculate CM achieved by your line.

Note: if a factory pays double overtime rate, and if there are incentives schemes, these are direct labor cost. Include these additional amount in the daily expenses.

Why you need this data daily?

In a production line, every day you are spending money (dollars) in workers' salary. You should know how much dollars you are earning on daily basis against the dollar you are paying to your workers. You also need to check the progress whether the earning is increasing or not after loading a new style in a line.

That is why you need to measure CM achieved on daily basis.

Industrial Engineer's Digest

How to Calculate Garment Production Cost from SMV and Monthly Salary?

Question: Some of the garment factories calculate CM (cost of manufacturing) of a garment by the machine SMV multiply with earnings per minute. And this earning per minute calculated based on monthly factory cost divided by monthly produced minute. Is this calculation logical? ... asked by an OCS reader

Yes, in the apparel industry, many factories use the method that you mentioned in your question for calculating the direct cost of production. From garment SMV and minute cost data, direct labor cost is calculated.

This is logical, provided the standard minutes (SMV) of operations are accurate and cost per minute is derived from a reasonable monthly salary amount (maybe minimum wages/higher than minimum wages).

Further, you need to incorporate line efficiency for deriving the labor cost. The monthly salary of direct manpower may vary depending on the employee's skill and grading. You also need to adjust the combined daily earning of an operator, to make it a satisfactory amount.

When you talk about production cost, either it is an estimated production cost or the actual garment production cost. In the subject line, production cost means the **estimated production cost** before you make the style.

The actual production cost per day and the actual cost incurred in a style can be calculated from the real employee attendance data (number of manpower involved in making the style) and their daily wages.

On the other hand, when we need to estimate production cost (direct labor cost) for the whole order, it cannot be calculated directly as we do not know how much manpower will be involved making the order. Secondly, the salary will be different for different employees.

Most importantly, we cannot guess how many garments they will produce daily and how many days will be consumed to complete the order.

In such a situation, the best way of estimating garment production cost is following standard time, wages per minute and benchmark efficiency. Where the wages per minute can be derived from monthly wages. Let's say, monthly operators' salary Rs. 10000.00. Per minutes wage is (10000/(26x8x60)) =Rs 0.80 (26 days work in a month).

Formula for calculating estimated labor cost per garment is shown in the following box.

> Estimated labor cost per garment = (Garment SAM x Labor wages per minute) / Line Efficiency%

Table-1: Calculation of labor cost per garment

Product type	Garment SAM	Labor wage per minute (INR)	Line Efficiency%	Labor cost/garment (INR)
Shirts	23	0.80	65	28.30
T-shirts	6	0.80	80	6.00
Polo	15	0.80	70	17.14

With the above formula, you calculate labor cost per garment (item wise). Once you know the estimated labor cost per garment, you can calculate estimated total production cost of an order. Production cost for the whole order = (Labor cost per garment x Order Quantity)

The second method of calculating production cost per garment from SMV and monthly wages.

Step-1: Calculate total monthly wages paid to the direct workers making the garment. Let's say 100 workers working in the sewing floor and total salary paid to them is INR 800,000.00

Step-2: Calculate total SAMs earned by the whole factory (sewing department only). The factory produced 30000 pieces of garments (including all styles) in that month. And earned minutes 1000,000. From this calculate cost per minute = INR 800,000.00/1000,000 = INR 0.80

Once you have the cost per minute of factory for direct labors, you can calculate production cost using this formula.

CM of a garment = (Garment SMV x Labor cost per minute)

PS: Labor cost per minute and labor wages per minute are two different parameters.

How to Do Production Scheduling Without SAM?

Question: *In the apparel manufacturing countries, most of the factories do not calculate garment SAM. So, is there any better way to make Production Plan without SAM calculation? ... asked by an OCS reader*

In the global garment industry, mass production is growing and at the same time technology and supporting departments are getting in place. But the real fact is - still most of the garment manufacturing companies do not have industrial engineering department. It is not only in Bangladesh; this is true even in India and other countries. And there are many companies that are managed without engineers.

Yes, there is a way to perform production planning tasks without garment SAM value.

The alternative way is calculating machine productivity to estimate line capacity and factory capacity. To measure machine productivity of a line, you do not need an engineer. If one knows the procedure and formula of calculating productivity of the production floor, can calculate machine productivity easily. Later this productivity data is considered as base of production planning.

Method of calculating machine productivity:

Machine productivity is defined as number of units produced per machine per day. From the daily production volume and machine used to produce those garments you can measure machine productivity per day. Formula used to calculate machine productivity of a line is shown here.

> Machine productivity = (Total production/No. of machines)

Assume that a sewing line of 35 machines has produced 280 pieces of garments per day (8 hours shift day). Machine productivity of the line
= 280/35 pieces per machine per 8 hours day
= 8 pieces per machine per 8 hours day

Finding daily line output data and machine numbers would not be a difficult job to anyone. Just for your information - style to style productivity will vary depending on work content of the style. So, you can prepare a database for machine productivity data of the previously produced styles. You can use historical data while required and update your machine productivity database after each style gets over.

How to use productivity data in Production Planning?

To explain this, I need to repeat few things here. Using productivity figure, you can measure following things those come under production planning functions

Factory capacity calculation:

Suppose machine productivity of your factory for ladies' blouse is 6 pieces per machine per 8 hours shift and you have total 400 machines in your factory. So daily production capacity of the factory would be equal to (400 x 6 pieces) 2400 pieces per day (8 hours shift).

Factory capacity calculation formula of a given product

> Factory capacity (in pieces) = (Machine productivity x No. of running machines in your factory)

Other planning related tasks that can be performed using factory capacity and line capacity figures are as follows -
- Lead time calculation
- Order booking
- Order scheduling
- Garment cost per piece

You can also measure line capacity using labor productivity data. But the machine number in a factory is more stable than manpower (labor). I would suggest you use machine productivity in measuring your factory capacity in term of how many pieces factory can produce in a certain period.

How to Improve Line Efficiency in a Piece Rate Factory?

Most of us believe that piece-rate operators are most efficient in the sewn product industry. As machine operators are paid based on the quantity they produce, they will put maximum effort into their work. Hence, one can expect the best efficiency of the factory having piece-rate operators. But this is not always true. Forget about the operator's maximum effort to produce standard hours (SAH) for a moment. And think on other factors those have a direct effect on the line efficiency.

Line efficiency depends on a couple of factors other than the operators' efforts. Here are few of those parameters -

- Skill level of the operators
- Line balancing
- Worker motivation

To improve the line efficiency (or individual operator efficiency) you need to work on the above points.

1. Skill Training for low performing operators

Working on piece rate does not mean that all operators will be working at 100% efficiency. There must be opportunity to improve employees' skill level. Work on training your operators. Use the following method to improve the operator's skill level.

- Measure the operator's individual efficiency
- Make a list of low performing operators (i.e. operators working at less than 50% efficiency can be considered as low performer).
- Observe your operators while they are working and study their method of performing jobs and movements.
- Check what are the unnecessary motions/movements you found operators are using at the time of sewing.
- Train them correct motions and give them time to practice.

Once you improve the operator's sewing skills, you will see improvement in line efficiency. If you do not know how to train sewing skills to your operators, you can hire an expert.

Here, I would like to share Paul's story on skill improvement for low performing experienced operators. This story will explain - through training you can bring a lot of improvement in operators earning and line performance.

*****

"Paul Collyer was working as a trainer in a garment factory, in the UK. He trained one girl on how to train operators and build sewing skills. Once she was trained completely, Paul sent her to the sewing floor to find out low performing operators and improve their sewing skills through training.

In that factory, most of the sewing operators were working for many years. So, they were experienced in their job. Operators were paid based on the number of pieces they were producing. The factory was running a piece-rate system.

As instructed by Paul, the young lady went to every operator and studied the operator's performance. She set a method to train low performing operators. When she spoke to experienced operators for the training, they ignored her. Few so-called experienced operators used to tell her that they are working for 15 years and they know better than her. They were in the age of her mother. They do not need to learn anything, at least from the young trainer.

The young lady became upset and reported back to Paul. Paul knew this would happen. He told the young lady, "don't worry, go to operators who have recently joined and improved their skills". She did the same.

After several weeks of training, newcomers started earning more than the experienced operators. Experienced operators were not able to earn as much as newcomers were earning with their full effort. All the experienced operators got surprised and they acknowledged the benefit of skill training.

All experienced operators who earlier ignored for skill training met Paul and requested him to send the young trainer to train them better skills so that they can also earn more money."

*****

I hope now you understand the power of training. There is always scope for improvement. The same way you can improve your line efficiency in your piece rate factory.

2. Work equalization or balancing of the line.

To produce more pieces, operators must have feeding full time so that they can continuously work and make garments. In a piece-rate factory you may have unbalanced WIP. To increase the line efficiency, you need to increase produced SAH using the same number of employees. By balancing the line, you can increase produced SAH.

Work on improving line balancing to improve line efficiency.

Refer to the earlier article on the line balancing - *How to Balance a Traditional Sewing line?*

3. Performance Incentive

By providing incentives to the employees, you can improve line efficiency of a piece-rate factory.

In piece rate production system, it is considered that operators are self-motivated in term of their earnings. But sometimes operators do not know their hidden skills and capability of earning more than the current earnings. Implement performance-based incentive for piece-rate employees also in combination with skill training.

You can set differential piece rates for a different level of individual operator efficiency. You can also link performance-based incentive based on the line efficiency instead of individual operator efficiency.

Follow these three methods for increasing line efficiency in your factory. These methods can be applied to salaried factories as well.

How to Plan Daily Line Output from Garment SAM?

Industrial engineers and production managers often need to calculate daily target production of a line. So, they need to plan daily line output.

If you are not calculating daily estimated line output, read this article. You can estimate daily line output if you have knowledge of a few required data. The scientific way of planning daily production (line output) is to calculate estimated production from garment SAM.

Daily Line output = (Available minutes @ Target Line Efficiency/ Garment SAM)

We need information for the following parameters.
- Garment SAM
- Day of production
- Learning curve efficiency

Garment SAM:
In the question it is mentioned that your aim is to calculate line output from garment SAM. So, first you need calculate garment SAM. You can get that in the operation bulletin of the style.

Day of production:
In the practice, after loading the style in a line, you cannot expect pick production on the first day of the production loading. In case you want to plan for the most appropriate production figure, you need to know the day of production. I mean, the number of day after loading of the line.

Learning Curve efficiency:
Due to learning curve on the initial days of line loading, line efficiency will be low compared to average line efficiency. If you just want to plan average production, consider average line efficiency%.

In case you do not have learning curve efficiency, you can use average line efficiency. Use line efficiency for the production day according to the learning curve.

- No. of machine (manpower) allocated to the line
- Daily working hours
- The formula for the estimated production (daily)

> Production Quantity= (No. of operators × Working hours per day × 60 × Average line Efficiency%)/ Garment SAM

Example: Assume that you are making a style of 20 minutes SAM. And details of the above listed parameters are as the following
- Day of production: 4
- Line Efficiency on day 4: 46%
- No. of operators: 32
- Daily working hours: 10 hours

Therefore, estimated line output will be
= (32 × 10 × 60 × 46%)/20 pieces
= 441.6 pieces or 442 pieces

How to make the calculation easy?

You can make the above calculation easily by using the following steps. You do not need to work on excel sheet to find the planned production for the day. Just use the following steps and get the production figure.

1. Calculate total minutes per day per operator - 600 minutes (assume 10 hours a day working)
2. Calculate production per operator @ 100% - 600/20 pieces = 30 pieces
3. Calculate production per operator @ actual line efficiency - 30 × 50% (assume avg. line eff. is 50%) = 15 pieces
4. Calculate production by the line of 32 operators – (15 pieces/operator) × 32 = 480 Pieces.

Use this method when you need to calculate daily line output from garment SAM.

How to Establish Thread Consumption Ratio

There is a formula for determining thread consumption for apparel products. In that formula, you will get multiplying factors according to machine type and stitch class.

To determine thread consumption, you just need to multiply seam length with those factors. This way one can estimate the total thread requirement for making a garment.

Actual thread consumption for a unit length of seam depends on the following factors.

- Stitch Classes
- Stitches per inches (SPI)
- The thickness of the seam (fabric thickness)
- Thread tension
- Thread count (thickness of sewing thread)

So, you are suggested to calculate garment thread consumption by your own. You can develop a multiplying factor according to your product categories and requirement using the following steps.

Step-1: Arrange material for testing - To calculate thread ratio, you need a sewing machine, fabric and sewing thread that will be used for bulk production. For fabric and thread, you can take similar thickness and same thread count in case actual is not available.

Step-2: Sew the swatch and measure length - Sew a seam of 12 cm long and take 10 cm seam out of it by trimming 1 cm from both edges.

Step 3: Calculate thread consumption ratio - Unravel both needle thread and bobbin thread from the seam. Remove crimp from the unraveled thread and measure its length in cm. Generally, it will be higher than the seam length. Now find the multiplying factor by dividing thread length with seam length. Assume unraveled needle thread length is 12.5 cm then needle thread multiplying factor is 12.5/10 or 1.25.

Using this method, you can calculate thread consumption factors of other sewing machines.

Step 4: Calculate thread consumption of a garment - Once you have consumption factors, it would be easy calculating total thread consumption. Measure the seam length of all operations of the garment and get thread requirement by multiplying thread consumption factor. Add thread wastage 5% for the thread that trims out from each end of the seam. (Note: wastage% may be different in your factory and for your products).

For quick reference, you can follow the following thread consumption ratios. Source: Coats thread consumption guide.

Stitch Class	Description	Total thread usage (cm per cm of seam)	Number of needles	Percentage of Needle Threads	Percentage of Looper /Under Threads (Incl. cover)
301	Lockstitch	2.5	1	50	50
101	Chainstitch	4.0	1	100	0
401	2-Thread Chainstitch	5.5	1	25	75
304	Zig-zag Lockstitch	7.0	1	50	50
503	2-Thread Overedge stitch	12.0	1	55	45
504	3-Thread Overedge Stitch	14.0	1	20	80
512	4-Thread Mock-safety Stitch	18.0	2	25	75
516	5-Thread Safety Stitch	20.0	2	20	80
406	3-Thread Covering Stitch	18.0	2	30	70
602	4-Thread Covering Stitch	25.0	2	20	80
605	5-Thread Covering Stitch	28.0	3	30	70

Note: the above ratio is arrived at with a stitch density 7 stitches per cm (18 stitches per inch). Changes in stitch density will change the ratio marginally.

If you want to refer ready-made thread consumption factors, and thread consumption for various apparel items, refer to Coats and A&E's Technical bulletins. You can download those bulletins using following URLs.
- https://www.coats.com/en/Industries/Apparel/Information-Hub (accessed on Nov. 09, 2020)
- http://www.amefird.com/wp-content/uploads/2012/09/Estimating-Thread-Consumption-.pdf (accessed on Nov. 09, 2020)

How to Calculate Thread Consumption of a Garment?

In the previous article, we have discussed thread consumption ratio. So, you already know the thread consumption ratios for different sewing machines. Now using an Excel template, you can calculate thread consumption of an apparel item. Use the following steps to calculate thread requirement of a given garment sample. We will take a t-shirt for calculating thread average.

Step-1: Take a garment. List down operations of the garment on your notebook or on a paper. You can refer the operation bulletin for list of operations if you have already prepared an OB for the sample garment.

Step-2: Measure seam lengths of each operations (in centimeters) and note what stitch class/machine type is used for stitching the seam. Like, Single Needle Lock Stitch, Chain Stitch or 3-Thread covering stitch (Over lock). Use the Excel template that is shown in the next page for thread calculation.

Add 10 cm for each seam as thread tails in two ends of the seam. (Note: This allowance figure may vary depending on machine type and order volume. If you already have the thread allowances for different stitch classes, you can use that data).

Step-3. Sum up seam lengths of similar machines together. See the example in the following Table-1.

Step-4. Multiply total seam length by thread consumption ratio (refer thread ratio in the previous article). Add thread wastage percentage to calculate total thread requirement for each stitch class.

Step-5. Finally calculate sum of thread consumption of different stitch classes to calculate total thread requirement per garment. Divide total length by 100 to get thread consumption per garment in meters (unit).

In an order, you need to make garments for different sizes and seam length for different garment size will be different. But for thread calculation, normally the middle size among the sizes can be considered for calculating average thread consumption per garment for a given order.

An example: Assume that you are given a crew neck t-shirt to calculate thread consumption. The operation breakdown and seam length given in the following table for your reference.

Follow the thread ratio provided in the previous article. You can use following matrix for thread consumption.

Table-1: Thread consumption calculation template

Operations	Seam length on following machine types / stitch classes (in centimeters)		
	SNLS	3 TH O/L	Flatlock
Ratio	2.5	14	18
Make neck rib	15		
Shoulder attach x 2		48	
Sleeve Hem x 2			100
Attach Sleeve x 2		120	
Attach Neck rib to Neck		76	
Sew neck tape with Main label	50		
Top stitch on Neck Rib	46		
Side seam and under arm sew		148	
Bottom Hem			130
Sub-total (Machine wise)	111	392	230
Thread consumption (M/c wise)	277.5	5488	4140
Add Wastage%	10%	15%	15%
Thread consumption with wastage (M/c wise)	305.25	6311.2	4761
Thread consumption per garment	In cm	11377.45	
	In Meters	113.77	

Exercise: Take one shirt item and one polo shirt from your closet and calculate thread consumption of both garments using above method.

How to Justify the Piece Rate of a Garment?

You know, many garment companies calculate workers payment based on the operation wise piece rate. This payment system is widely used across the globe.

Secondly, this is a very simple way of calculating workers payment once the piece-rate is agreed.

Garment companies decide the piece-rate per garment per operation after the style is loaded to a production line.

My question is, how do you justify the garment piece-rate that is agreed between the worker and the company management?

It is true that workers are motivated by the piece-rate payment system. Operators earning is directly related to their production quantity. The factory management also gets confidence that workers will work at their maximum speed and effort to earn more money. Factory owners feel satisfied that their workers are motivated.

How is the piece-rate derived for a stitching operation?

In most cases, piece rate is finalized after a negotiation with operators, contractor and calculate daily earning capability of their operators (Number of pieces stitched X Piece-rate per garment). Normally, operation rates are finalized within 2-3 days of production start.

The factory needs to ensure that they give a whole day's salary to their piece-rate workers. When the piece-rate is set, the factory considers - the daily minimum salary to earn by operators, order quantity in a style, operation handling difficulty, and handling of a special machine.

A factory studies garment operations and operator's production rate in initial 1-2 days. Factory assesses operator's production rate in pieces. How many pieces they can produce when they are provided with an uninterrupted feeding of cuttings (bundles). According to that operation-wise piece-rate are derived.

For example, if you need to provide minimum daily earning equivalent to ₹300 and if an operator can make 600 pieces (in one operation) in a day, piece-rate for that operation would be ₹0.50 (300/600). This formula is used as a benchmark of the Piece-rate.

How to estimate possible piece rate for an operation without having a work-study team?

The piece-rate amount can be justified by using capacity study and operation cycle time. Here is the guide to justify whether your price rate of garment operations is higher or less. This guide is specially written for companies those do not assign Standard Time (SAM) to the garment operations and do not have industrial engineering team.

1. Let the operator work for one full day in the operation for a new style. If you think, an operator cannot reach normal speed in one day. If needed, you can give them two days for learning.
2. Measure the average cycle time of the operation (take at least 10 consecutive operation cycles' time). Use a template for capturing cycle time. If you have more than one operators working in the same operation, normally their production rate will be different. So, measure the cycle time of all operators and consider the lowest cycle time for the operation for calculating piece-rate.
3. From the cycle time calculate average operation cycle time
4. Add 15% allowances on the average cycle time and calculate the given time for operations and convert it in minutes. (This allowance in machine delay, fatigues, and personal need).
5. Next, calculate per minute rate from the daily minimum wages of a skilled operator. Per minute salary rate = (Daily wages/480 minutes). (Standard daily wages are set in 8 hours work). In the example, the daily wage is considered as ₹ 300.
6. Now calculate operation rate. Piece-rate = (Per minute salary x average cycle time). This is achievable piece-rate for the given operation.
7. Now you have two data - The agreed piece-rate and calculated piece-rate derived from the average operation cycle time. Compare the two rates - and you will find which one is higher.
Use this as a measuring tool for garment piece-rate assessment.

See the following example of 3 operations. In the following table, agreed piece-rate is named as Given Piece-rate. You can see the difference between the two rates. If the variation is very high, you may decide on revising the piece-rate. The average operation cycle time shows the potential production capacity of garment production.

Table-1: Cycle time data collection format + Piece-rate calculation

Operation name	Operator name	Operation cycle time (in seconds) based on study					Avg. cycle time	Avg. time with 15% allowance	Avg. time in minutes	Daily Wages	Calculated piece-rate	Given piece-rate
Back seat join	Ashok	20	24	19	21	22	21.2	24.38	0.41	300	0.25	0.20
Pocket overlock	Shanthi	36	45	42	43	40	41.2	47.38	0.70	300	0.49	0.25
Side seam raw stitch	Nadeem	30	36	35	34	40	35.0	40.25	0.67	300	0.42	0.50

By following the above method, you can justify the piece-rate of garment operations given to the piece-rate operators by a factory.

Industrial Engineer's Digest

How to Encourage Workers in a Garment Factory?

Question: *"We have now about 280 operators in our factory and we want to apply production bonus, please note that the collection bonus is not working because they keep changing operators from one line to another line to replace the absence. Is there any method to apply to encourage workers?"*

Yes, you can apply some other methods than the collection bonus (group incentive system) to encourage your workers. I will explain some of those in this article.

Money is the topmost thing that can motivate and encourage shop floor workers. But yes, there are other kinds of benefits that can also motivate your workers. If not all, most of them can be motivated.

1. Group performance bonus

I can understand the issue you are currently facing in the group bonus (line wise) system for your shop floor workers. But it is a common practice that we need to move workers from one line to another to manage the production line and keep production going.

There are some reasons for shifting workers from one line to another. It may be employee absenteeism (as you mentioned in your question), for better line balancing, or for better resource utilization, for new style set-up etc. Though workers move from one line to another line and work with a different group of workers, group incentive can be calculated, and the incentive amount can be provided to the workers. For such cases, you need to calculate the group incentive amount on daily basis.

There are two conditions you need to follow - If most of the employees work in the same line in a day, you can start a group bonus system. For this, you must have a good data capturing system. You can still distribute the performance bonus amount on weekly basis after accumulating daily bonus amount.

I have seen garment factories providing group incentive as well as an individual incentive in a similar production environment.

2. Individual performance bonus

If you really find it difficult managing and calculating incentives amount for group bonus, when workers work in a different line, go for the individual performance bonus. Design an incentive scheme for providing a bonus to the individual operators based on their individual performance. Say overall efficiency% of the workers. You can set eligibility criteria for earning the performance bonus, like

- Individual worker's overall efficiency should be 60% to get the jump bonus after that every point of efficiency increment, they will earn an extra amount.
- Line efficiency should be achieved x% to be eligible for the individual bonus.
- Weekly or monthly attendance should be 100% or one leave per month for getting the bonus.
- A few more criteria can be included when giving the bonus.

(Note: The Above bonus eligible criteria are shown for your reference. This may not work for your factory. You need to set your own bonus eligible criteria)

3. Employee attendance bonus

To reduce the employee attendance issue that you have mentioned in your note, you can provide your workers attendance bonus. The attendance bonus should be lucrative too to the workers. This will gradually improve the employee attendance in your factory. As a result, employee movement between the production lines will come down.

4. Skill training of employees to enhance individual performance

Some of your workers may need skill training to improve their performance on the job they used to do. By increasing their performance, they will be eligible for earning more money as their skill level will be upgraded. Even if you provide an individual performance bonus, they will get a chance to earn extra money.

5. Overtime working and double rate for OT work

I will still focus on creating possibilities of extra earning opportunity for operators. If you allow overtime work, which will allow operators earning extra dollars, many operators will be interested on working with you. You should provide an overtime rate at 1.5-2 times than the regular hour rate. In another article, I have discussed on overtime working – benefits and negative impacts.

6. Good HR policies

Though I said money can motivate workers, there are many other ways to make your workers and encourage them. There are many small things can be considered under HR policies those can motivate workers to stay with your company and give more effort to achieve the factory's best performance. For some example,

- Allowing leave to workers when they need it
- Provide low cost and quality food inside the factory pantry
- Paying the salary within the defined date
- Establishing a good work culture inside the factory and supporting co-workers and supervisors.
- Taking care of workers issues

Introducing the performance bonus is a good way of motivating workers. But you should have a strong system to measure and calculating employee performance and all eligible workers should get the incentive amount as you agreed to provide them. Incentive scheme must be easy to understand and transparent to workers. By encouraging and motivating your workers you can produce more garments with the same number of workers and staffs.

Sewing Operator Recruiting Test Procedure

Question: *How to test an operator in woven industry? I want to take test of an operator before hiring him. What will be that test? ...asked by an OCS reader.*

Normally, garment factories follow standard procedures for hiring sewing operators. I have seen it in many garment factories that industrial engineering department is responsible for conducting on-the-job test for candidates to ensure whether candidates are qualified or not as sewing operator.

In case factory does not have IE department, line supervisors carry out this test. Under the recruitment test, sewing operators need to go through few tests and show their expertise in the given tests to be get hired.

When you are hiring an operator and want to place her directly to a line, she must be a trained operator and must have worked earlier in another organization. For such a trained operator, you need to test her sewing skills on following parameters –

1. Efficiency level in different operations (choose operations of the styles that you make)
2. Quality of the seam and stitches made by the candidate
3. Types of machines she can handle
4. Types of fabric she can handle (like knits, woven or light fabric, thicker fabric)

These tests also help you to prepare skill matrix of these new recruits. You can add their skills in the existing skill matrix (if you have one).

Operator recruitment test procedure
Factories normally follow the below procedures in operator recruiting test.

1. Ask candidates to make mocks of the garment components for the product you make in your factory. For example, if you are making shirt then give candidates to make Collar, Cuff, front placket, or Chest pocket making and attaching to front panel.

2. Assess their work on the mock in terms of quality, performance level on the above parameters. If they make mock-ups up to your satisfaction level, you pass them for sewing skills. Then forward them to HR team for HR verification and recruiting formalities. If they do not meet your satisfaction level, you can reject them.

3. At time of making mocks, also check how the candidates handle a machine. Like,
- how they do threading a sewing machine,
- needle attaching to machine,
- feeding of bobbin into shuttle,
- how fast they run a machine etc.

From this test method, you can assess how good your candidates are. From the test result you can decide who all are suitable and qualified for your requirement.

Further you can observe candidates' learning attitude, body language, understanding of your instructions, how they talk etc. Based on your requirement, you can select one who meets your requirement.

Part - IV

Improve Factory Performance

Skill Upgradation of Sewing Machine Operators

Skill upgrading of an operator can be done in various ways. It may be building skill in different machines – one who can handle single needle machine, you can train him to handle an overlock machine, building skill from single operation to multiple operations. Building skill from easy operation to make a difficult operations – one who is skilled in straight stitch, train him to make a curve stitch. Skills to be imparted to a machine operator depends on the demand of a sewing line and product constructions.

Skill upgrading also means improving performance level of low performing sewing operators. Experts say, certain percentage of sewing operators among the operator pool, work at the low efficiency level. They are called as low-skilled operator. Some of the operators are low-skilled because due to below reasons. They -

- work at slower speed,
- run machine slowly
- Do not use correct methods or/and movements to perform the jobs
- Operators are not trained by professional trainer (a trained trainer). So, they do not know about good movements and bad movements. They do not know exactly how they can do the same job with less effort and achieve a better performance by using better methods.
- Operators are not comfortable or habituate with the job/task they are performing.
- Not confident doing a job or/and running a machine and hesitate during performing a job.
- Fear of making faulty (defective) work when machine run at a higher speed

When it comes for upgrading skill for machine operator, it does not mean that only low skilled operators can be upgraded. Everyone's skills can be upgraded from his/her present skill level as explained above various form of skill upgrading.

When you are in garment manufacturing business and you want a continuous growth, you must build operator's skill through training. It will help you to increase manpower resource utilization. In this article I will explain the ways to upgrade operators' skill in their sewing task (performance level in the task what they already performing). Use of scientific training method is necessary for operator training. To start with upgrading existing operators, employ a trained sewing operator trainer who can train low skilled operators.

Remember, trainer must follow a systematic training course for the sewing machine trainees.

Following steps will give you an idea how to upgrade your existing machine operators.

Step-1: Selection of Low skilled operators:

Trainer needs to find low-skilled operators from the floor based on the criteria for low-skilled machine operators. For example, lower individual efficiency than line average efficiency or target efficiency.

Step-2: Observation of Movements:

Trainer needs to look into/observe the movements of a low skilled/low performer. Study each movement an operator performs when operator sew garments. Do cycle timing if required to understand more.

Step-3: Movement analysis:

After motion study analyze what operator is lacking at the time of sewing a garment, what all things are not correct (excess movements, handling of work, machine speed, rhythm on work, concentrate on the job or not etc.)

Step-4: Training on Good Movements:

Teach an operator the correct of way of doing the job what he is doing at present. It is all about movements. How to pick-up parts, from where parts to be picked up, where to dispose, how to dispose, using both hands, simultaneous task by two hands, how to move/slide part under needle, correct sitting posture etc. Teach correct movement. Each time trainer teaches a new movement, operator must be given time for practicing. Also measure time when operators practice. It will give trainer to see how each of the low performer are learning and improving their performance level.

Step-5: Training on Speed:

Teaching of correct movements and methods are necessary but at the same time operators need to be trained to run machine at higher speed.

Operators who run machine at slow speed teach them how to run machine higher speed without compromising quality.

Step-6: Think out of the box:

It is not always the operator that restricts in reaching operator's efficiency at certain level. You must think other ways to make performance improvement in a specific operation with the same operator. For example, guides, attachments, or a fixture can help to improve performance of an operator.

Can an Operator Hit 100% Efficiency?

If you ask me this question, my answer would be "Yes, an operator can hit 100% efficiency".

Some of your operators can hit 100% efficiency. Provided that the operator has been given with enough work, good workstation layout and a good machine. You know the efficiency calculation formula. For calculating employee efficiency, you need a standard time of the task done by an operator.

You may be thinking, how that is possible. Let me explain it using the parameter used for establishing SAM.

You may know how standard minute of a sewing operation is calculated. Let us say the standard time (SAM) is established using a PMTS software. If parameters are set right, then you will find few operators hit 100% efficiency.

Eight main parameters are considered in calculating garment SAM. Before questioning your operator, you must choose correct options from each of the following parameters to establish accurate SAM of an operation.

1. Machine rpm – Do you provide your operators same machine you used during SAM calculation?

2. Stitch per inch (SPI) – whether the SPI used in production is same as used in SAM calculation?

3. Stopping Accuracy – what is the right option for stopping accuracy for the operation?

4. The difficulty level in material handling – do you consider the right option for material handling?

5. Seam length – what you do when multiple sizes run on the line? Single SAM for all sizes, right?

6. The number of stopping (burst) in a single seam – you may use the standard number for burst for a seam length provided by a software provider. But you need to think - does standard stopping formula works for each operation and is it same for each employee?

7. Machine allowance as per machine type – you may be right with this.

8. Contingency allowances – you may follow standard allowance percentage.

Where you need to consider all the above parameter to establish the right SAM, there is a chance of making mistake in calculating SAM. Anyway, your SAM will be loose or tight.

Now assume that you have established SAM which is moderate for an operation (not loose, not tight). A highly skilled operator can beat your SAM by following ways.

1. Taking less stopping (burst) than you considered in the SAM estimation. In each burst, an operator can save 17 TMU
2. Utilize less allowance (machine and contingency) than the allowances added to the SAM. Approx. 15-22% allowances added to SAM. But many time, operator does not need that much allowance.
3. Operator running machine at a higher speed (RPM) than you considered it in SAM calculation.
4. Operator uses less SPI when sewing a garment compared to the standard SPI used in SAM estimation.
5. Sewing a seam of shorter length than a length used in SAM calculation

Let's say, SAM of an operation is 1 minute, for a 10 pieces bundle, standard time would be 10 minutes. Assume that by applying the above tactics an operator makes each garment in 0.8 minutes. So, bundle completion time will be 8 minutes (for 10 pieces). In this example, operator produced 10 minutes in 8 minutes time.

Therefore, operator's efficiency will be = (10/8) x100%= 125%
Your operator hit the 100% efficiency.

Following up the Hourly Production Target

To control the line performance and daily production, line supervisors need to follow up production in regular intervals and then compare it against the hourly target. If a supervisor wants to know what the line output per hour is then it is easy. He can count pieces physically or ask a helper to count stitched pieces and note it down in the hourly production report in a white board. You know the production target of the lines and you need to achieve it.

The actual control can be done only by taking hourly production figure of each operator. There are various ways to know operator wise actual production in the last hour or from morning to till hour.

In most of the factories production target follow up is done at the last operations of the style. For hourly or bi-hourly production data, factory can employ a work-study boy who can collect production from each operator. He can count pieces and note it in the operator's job card. Or operators can note down their production when they pass the completed bundle to the next operators. After collecting production data from the line, hourly line output is displayed in the hourly production board.

In this system, there is a delay for an hour or more. When a supervisor gets the report, most of the operators may produce more garments. For example, daily production report is prepared in the next day morning in most of the factories.

The second way of capturing hourly production is through installing Bar Code tracking system. By using this system, operators stick barcode stickers on a gum sheet as they complete the bundles. Then one person scans those bundle tickets (Barcodes) and get hourly production of each employee in a computer screen.

Another way of getting real-time production data for supervisor is RFID based real-time production tracking system. This system can help line supervisors to track movement of bundles in a line. WIP level at each workstation. You will have operator wise and operation wise production quantity at any point of time.

A real-time system can generate hourly production report and line balancing chart for each line. Through line balancing chart (graph), a supervisor can do hourly production follow up. And they can take data-based decision to control production.

20 Ways to Improve Shop Floor Productivity

Higher productivity brings higher margin in a business. And an increment in productivity reduces garment manufacturing cost. Hence a factory can make more profit through productivity improvement.

In this article, 20 ways of productivity improvement are discussed that will certainly help factories to boost up current labor productivity. These productivity improvement methods can be implemented by all as these are doable. Machine productivity as well as labor productivity increases when a factory produces more pieces by the existing resources (Manpower, time, and machinery).

When I investigate the processes and operations during my visit to factories, I find improvement potential is there in every area. Initially, you may not be able to find and measure potential areas. Still, you can improve productivity by applying some of the following steps.

I have mentioned 20 ways where you can focus and start working on improving productivity. Most of the tips mentioned in this article are mainly on time-saving tips, discipline, and proper planning. To get an excellent result you may need external recommendation and support but without the external guidance, you can surely get measurable improvement once you start your journey.

One thing you need to understand that the application of the following steps will vary based on the production systems. Here are the 20 different ways to improve productivity in a garment factory.

1. Conducting motion study and correcting faulty motions
2. Hourly operator capacity checks
3. Conduct R&D for the garment
4. Use best possible line layout
5. Scientific workstation layout
6. Reduce line setting time
7. Improve line balancing
8. Use work aids, attachments, guides, correct pressure foots and folders
9. Continuous feeding to the sewing line
10. Feed fault free and precise cutting to the line
11. Training for Line Supervisors
12. Training to sewing operators
13. Setting individual operator target

14. Eliminate loss-time and off-standard time
15. Real-time shop floor data tracking system
16. Using an auto trimmer sewing machine (UBT)
17. Installing better equipment
18. Inline quality inspection at regular interval
19. Operator motivation
20. Plan for operator's Incentive scheme

All the above ways are briefly explained below with examples.

1. **Conducting motion study and correcting faulty motions:**

    There is a saying "Even the best can be improved". So, you need to visit the shop floor often and observe operator's working method and movements. Prepare a checklist for good methods and movements for sewing activities. At the time of **motion study,** observe operator's movement and compare with your checklist. If you found wrong movements is used by the operator or unnecessarily extra movement is present in the operation cycle correct it. If needed deskill operator. By doing this you can reduce operation cycle time and can improve labor productivity up to 100%* in individual operations (*in 20% of the total operations as per Pareto's 80-20 Principle).

2. **Hourly operator capacity checks:**

    Employ work-study personnel (if you do not have) and check operator's production capacity hourly or bi-hourly interval. Compare actual operator's hourly production with their potential capacity. If production is less, then question them why? It helps in two ways – first, when the operator's capacity is checked at the regular interval they will be under pressure. Secondly, work-study personnel start thinking on methods of how cycle time can be reduced.

3. **Conduct R&D for the garment:**

    A Research and Development (R&D) team in the factory brings a lot of benefits in term of preparation for the bulk production. R&D can be taken as a preparation stage for bulk production. This department makes samples and checks potentially critical operations. They plan for the requirement of special equipment, advice changes in terms of construction without changing styling. E.g. if an operation contains some raw stitches, which does not affect the final look of the garment, then that operation can be avoided to save time. They plan for skill requirement for the operations. As a result, production runs without any break or with less no. of breaks. As it reduces the chance of the production breaks in production for unnecessary reasons, line productivity does not come down.

4. Use best possible line layout:

Line layout means arranging of machines and center table in a line (trolley with wheel) as per style requirement. The main purpose of choosing a better layout is to reduce transportation time in the line as much as possible. A stable line is not a good idea if you produce multiple products in the same line. A straight assembly line with a center table at the left side is good for a product that has no preparatory work and individual operation SAM is nearby the pitch time. When a style includes a lot of preparatory work (for garment parts), it is better to make garment parts in sections and later assemble all components. If possible, use overhead transportation system.

5. Scientific workstation layout:

The workstation layout defines from where an operator will pick up the work (garment components) and where she will dispose stitched garment. A scientific layout is defined as minimum reach for picking up and dispose of components. Every components and tool (trimmer) must be kept within operator reach. During workstation designing, engineers should follow these key principles.

- Components to be worked on should be positioned as near to the needle as possible.
- The direction of the components where it positioned on the table or track should be such way that during moving component to the needle point does not need to turn it.
- Placing of work at the same plane of the machine table so that operator can easily slide it to needlepoint.

The purpose of designing a good workstation layout is to minimize the material handling time as much as possible. Thus, you can reduce operation cycle time. The secondary benefit of a good workstation is operators can work at the same pace without fatigue. When you design a workstation layout, do not forget to consider ergonomics.

6. Reduce line setting time:

It has been observed that a line reaches at its pick productivity level on day 6-7 after loading of an order. The time lost in the initial days (during learning curve) brings down the average labor productivity for the whole style. Reason - a lot of time is lost during setting of the line for a new style. This reduces overall machine productivity and line efficiency. So, to maintain line productivity level you need to work on minimizing line setting time.

To reduce the line setting time, engineers must study the garment thoroughly, prepare operation bulletin with machine requirement and machine layout plan prior to setting up the line.

Engineers need to coordinate with line supervisors and maintenance department with their plans and requirements. This will help supervisors and maintenance department to be pro-active in arranging required resources.

7. Improve line balancing:

Purpose of balancing a line is to reduce operator's **idle time** and maximize operator utilization. In a balanced line, work flows smoothly, and no time will be lost in waiting for work. At the time of line setting, select operators for the operations matching operator skill and skill required. Following this method, you will select highly skilled operators for higher work content operations. Once the line is set, conduct the capacity study at a regular interval.

Use pitch diagram method to find bottlenecks inside the line. You must think how you will minimize WIP level at bottleneck operations. Read the article written on the line balancing topic in this book. Once you start increasing operator utilization through line balancing you will get extra pieces from the same resources in a day.

8. Use work aids, attachments, guides, correct pressure foots and folders:

These are some kinds of time-saving devices that facilitate the operator to perform their work effectively with less effort. If work aids are used effectively, you can reduce operation cycle time from the existing cycle time of the operation. In new and small factories where there is no experienced technical person (maintenance, IE personnel or production manager), they are not aware about the usage and availability of work aids. So, their operators sew garment free hand.

Labor productivity is comparably higher for the factories that widely use work aids than those who do not use work aids for the similar products. Folders and attachments are also very helpful in producing consistent stitching quality. On the other hand, work aids, guides and fixtures reduce operator's movement and weightlifting. In my research study at NIFT, I had improved labor productivity up to 18.03% using work aids in some sewing operations.

9. Continuous feeding to the sewing line:

It is not a fault of production department if they do not get cuttings to sew. All plans and efforts towards productivity will fail if factory does able feed sewing lines continuously. "No feeding or irregular feeding" is one of the top reasons for lower productivity in a poorly managed factory. Poor production plan, wrong selection of product mix, fabric unavailability, and ineffective cutting department are some reasons that stop continuous feeding. Once operators get the rhythm, they should be given non-stop feeding until style changeover. If you know there is the unavailability of cutting in near future, then plan it accordingly.

10. Feed the line with fault-free and precise cuttings:

Stop cutting and trimming of extra fabric from cut components in the production floor by the sewing operators. If your cutter is not able to provide precise cutting, train them and improve your patterns. Avoid feeding faulty cutting to the lines.

When operators cut fabric, he performs the additional task in the operation cycle. If in some cases trimming of cut panels is intended, then that task must be included in total work content. Otherwise, you will get wrong (less) efficiency for the operator. Secondly, cuttings with fabric defects, and pattern problem are issued to sewing line. As a result, line produces defective garments. Alteration and repair work for defective garments reduces labor productivity.

11. Train line supervisors:

Line supervisors are shop floor managers and productivity controllers. So, each supervisor must be trained with fundamental management skills and communication skill. Still, in most of the supervisors in Indian garment factories are upgraded from tailors. Most of them do not acquire a technical qualification in supervising number of people. Their main job is providing instruction to operators and transferring information. For which communication skill training is required for supervisors.

Secondly, a supervisor should understand the fundamental of industrial engineering like operation bulletin, skill matrix, workstation layout, good movements, capacity study and theoretical line balancing etc. If they understood these, they could help engineers or work study boys to improve line performance. The supervisors' training can bring changes in managing and controlling production lines and will improve labor productivity.

12. Training to sewing operators:

Operators are main resources in the apparel manufacturing. They are the most valuable resource to the company. So, the factory must work on developing operator skill where required. "Training is not a cost but an investment" said by many experts. Production from an operator depends on the skill level to the task. A low-skilled operator will consume higher resources (time) and give less output.

You will find stitching quality related issues with low skilled and untrained operators. As the skill level of the operators is increased through training, lines output will improve. Training means a lot of time and money. Training should be given only on specific tasks that will be performed by the operator.

13. Setting individual operator target:

Instead of giving an equal target to all operators working in a line, give individual target as per the operator's skill level and capacity. Set an achievable target for each operator so that they would try to reach the target. This will help improving the operator's individual efficiency. Use tricks for increasing the target step by step. Take care of the operators who are under target. They may need skill training.

14. Eliminate and reduce non-productive time:

Utilize the operator's time as much as you can. There is no better alternative than just stopping operators sitting idle to improve operator productivity. Non-productive time such as waiting for work, machine breakdown, power failure, and repair work kill your productivity. Start reducing non-productive time as much as possible by taking required actions. To start work on this point, you must track off-standard or non-productive time data according to different categories. Once you have the analysis and Pareto of non-productive time you can think and plan on reducing it.

15. Real-time shop floor data tracking system:

For the continuous improvement and prompt action on failure, you need information from the shop floor as early as possible. Important information needed such as hourly production, line balancing, WIP, tracking bundles and quality performance of the line. If corrective action is not been taken early problem may increase as time goes. An RFID-based real-time shop floor production tracking systems can help you tracking shop floor information within a second.

16. Using an auto trimmer sewing machine (UBT):

Just think how many pieces an operator is producing in a day. Each time an operator trims thread using a trimmer or hand scissors consume time minimum 50 TMU or 2 Seconds (approximate). A rough estimate, in a day an operator will lose about 20 minutes just in thread cutting. In an operation of 0.5 SAM, an operator can make 40 extra pieces. Even a machine without auto trimmer consumes more sewing thread. For the betterment, those who still using heavy (half kilogram of weight) scissors for thread trimming, can start using hand trimmer.

17. Installing better equipment:

A low performing machine is not acceptable where some of your good machines are idle in the same building. Use the best of your resources. If sewing machines or equipment do not perform well, operator motivation goes down. Repetitive breakdown of machines increases the loss time and bring down overall line efficiency and labor productivity. I have seen lines where UBT machine is used in long seam operations and comparably lower work content. On the other hand, where shorter seams are being stitched, most of the time spent in thread trimming for taking out work from the needle, normal lock stitch sewing machine is used.

18. Inline quality inspection at regular interval:

Traffic light system is one of the effective garment inspection tool to reduce defect generation at source. Less the number of defects generated, less the time will be lost in repairing it. Inline checking system will alert operators in concentrating their job. It also helps in another way. At the start of the style, an operator may not understand the specification, interaction with quality inspector will make an operator clear about the quality requirement. Poorly managed factory loses productivity up to 10% due to repair and rejects as mentioned by Dr. Bheda in his article "Productivity in Apparel Manufacturing".

19. Operator motivation:

Operator's will be the most crucial part of productivity improvement. If they are motivated, they will give extra efforts into the work. Employee motivation generally depends on various factors like work culture, HR policies, bonus on extra effort for achieving the target. In garment manufacturing, operators are getting motivated through extra money (performance incentives). Operator motivation can be improved by sharing a certain percentage of your profit gained from the operator's extra effort.

20. Plan for operator's Incentive scheme:

Paul Collyer, a British expert says, "In British factories, in a non-incentive environment factory can reach up to 80% efficiency level and if the manager expects more than that they had to provide an incentive to operators as well as to the supporting team". If we look into Asian factories, in a non-incentive environment factories find it is very difficult to reach up to 40% efficiency. You can see the potential efficiency that can be converted into money.

You know employees come to work in your factory for money. Initially, you may think that an incentive scheme may reduce your profit. But it works in the opposite way, provided that the incentive system is fair for the workers and has been implemented correctly. I have seen factories where operator efficiency reaches up to 76% from 45% after implementation of an incentive scheme. An incentive scheme will give a lot of other benefits in return as a byproduct. An incentive scheme designed with multiple parameters may bring discipline on the shop floor. As operators give extra effort to the work, efficiency as well as productivity of the operator increases.

20 Tips to Increase Shop Floor Performance

There are many heavy-duty theories for improving production in the garment industry. Many of garment factories go for implementing such difficult-to-understand and difficult-to-apply methods. Somehow, we forget to apply simple things those have enough potential to bring improvement in the production.

Floor performance can be improved by saving time, increasing production, saving cost of production, increasing workers' health and safety etc. All these are possible by taking care of small things. To make your factory better and improve overall performance, keep your eyes on small things as well. Bring changes that give you greater benefits.

Following are the simple and easy to apply tips. Most of these tips are on *preparation* and *disciplines*. Practice and get BIG benefits from small changes in your daily work.

1. Do not accept incomplete production file from merchandising team to start production activities.
2. Prepare line loading plan well in advance to avoid waiting after production starts.
3. Balance lines to avoid bottlenecks - balance line as frequent as possible based on the actual production data.
4. Keep enough WIP to feed the line whole day.
5. Do not allow production lines to generate excess inventory. Excess WIP in a line may reduce line/batch performance.
6. Provide extra bobbin to operators. This simple practice saves time in refilling the empty bobbin.
7. Feed defect free cuttings to the production line.
8. Save time by quick replacement of broken needles and empty thread spools.
9. Only feed ready cuttings (component) to sewing lines. Do not allow re-cutting and trimming of components in the sewing line.
10. Use under-bed trimmer (UBT) sewing machines - it improves stitching quality, saves time and improves performance.
11. Keep floors and table tops neat and clean - this improves employee's productivity.
12. Do not keep any obstruction on the walking space in floors.
13. Design workstation layout considering motion economy and ergonomic.

14. Keep spare machines for quick replacement at the time of machine break-down.
15. Schedule regular machine maintenance and follow the schedule.
16. Use work aids, guides, and attachment wherever applicable.
17. Avoid parts matching in the line (by the sewing operator), if required keep one helper for part matching.
18. Capture hourly production data and make hourly production report, display hourly production data on front of the line. Visuals always give positive impact to workers.
19. Do not give back defective garments to the running line. It breaks working rhythm.
20. Do not distract operators from their work by talking to them unnecessary.
21. Bonus tip: Keep drinking water dispenser as near as possible to sewing lines.

Start implementing the above tips that you find easy to implement. After implementing some of these tips and executing it, you will surely find improvement in overall floor performance.

9 Ways to Increase Sewing Operator Efficiency

Operator efficiency is the performance measure of an operator. With the daily activities, you are also responsible for improving sewing operator performance and factory performance. You can reduce manufacturing cost by increasing sewing operator's efficiency. You know the apparel manufacturing sector is a labor-intensive industry. So, it is essential for working on skill and performance enhancement of the shop floor workers, especially sewing operators.

The purpose of increasing operator efficiency is to reduce labor cost per unit and increasing profit margin.

I have shared 9 ways you can follow in increasing operator efficiency. Here I have mentioned pointers only. For the detailed guide for increasing sewing operator efficiency, read previous article – 20 ways to improve productivity.

Tips for increasing sewing operator efficiency

1. Develop operator's sewing skills through training on the job. Train them on good movements, correct material handling and train better method of performing a job.
2. Motivate operators by providing incentive based on their performance (efficiency level). You can provide incentives based on the individual performance or a group bonus based on the performance achieved by the production line.
3. Assign operators to the tasks on what they are skilled. If operators are assigned to operations on which the operator is low skilled, they will work on less efficiency. If you don't have an alternative skilled operator for a job, train your existing operator first (point no. 1) to develop his/her skill level.
4. Improve the working methods where possible by motion study and motion analysis. Eliminate excess motion from the existing working method.
5. Do not forget to design a good workstation layout based on operation requirement. Reduce excess reach. A good presentation of work is also important.
6. Supply work continuously to operators. While an operator is assigned a work with less work content and he/she had idle time, give him/her one more job.

7. Avoid unnecessary interruption by supervisors, quality checkers and other things like defective pieces is supplied to operators.
8. Give operators an achievable production target. Record operator's hourly production and encourage or create pressure to operators for extra effort if they produce less than their capacity and given target.
9. Do not do much overtime. Excess overtime exhaust employees. It has been found that on the overtime working hours, employee's efficiency is low compared to the normal day efficiency. You must have one day weekly off in a week.

By applying and adopting the above means you can improve operator's efficiency from the existing efficiency level on the specific jobs. To see the improvement, you must measure the operator's existing efficiency and current efficiency (after implement of above mean/s).

Here, I have given a random list of things you can implement to increase individual employee performance. Start with one that is easy to implement and gives you a better result.

9 Things You Can Do to Control Production in a Factory

Controlling the shop floor means getting daily production as per target plan, producing the desired quality of garments, and the production cost and other factors are under control. Other production factors are like reduced line setting time, no excess loss-time, and no excess overtime etc. Utilization of resources and line performances can be improved when you have better control on your sewing lines.

Normally, line supervisors are responsible for controlling sewing lines. But when it comes to data analysis and data-based decision making, they are not trained on that part. So, industrial engineers' (IE) intervention is necessary in production control.

In most of the garment companies, IEs are given multiple tasks. Some of those are related to managing and controlling the sewing lines and rest are not related to production floor. Most of the IE managers are involved in management meetings, project implementation, and solving employees' issues. In true sense they cannot concentrate to their actual jobs. In such situation it is difficult to control the production.

In this situation, this is the right question for discussion - how to control the shop floor with busy work schedule.

You may already know about the primary jobs of an IE, at least in your present company. You are hired to control production cost and reduce manufacturing cost continuously or improving other processes which indirectly affect production cost and improving line performance by various means.

To achieve these objectives, you need to do multiple jobs. Some of the major tasks you must do daily basis are discussed below to control the sewing lines (sewing floor).

1. **Check operator attendance report:**

    You cannot keep observing every operator whether they are starting their machine on time in the morning. If operators come late, all minutes will be lost until operators start the machine and start sewing. In case someone is absent, you need to find an alternative for that absentee.

2. **Prepare hourly production report of all operators:**

    Most factories display hourly line output report in a whiteboard. That is not enough to control the sewing line. You need to know operation wise hourly production. In case individual production tracking is not possible, collect production data for each section of a line. Technology is available to perform this task automatically.

3. Analyze lost time (non-productive time)

The major reason of efficiency loss of a line is lost hours due to lost time and idle time. To reduce lost time, you need to measure those. In another article, I have covered the topic - non-productive time and common lost time categories.

4. Balance the production lines:

If you track hourly report for all operations in a line, it will be easy to understand work volume throughout the line. Make a line chart or bar graph. On the graph, you can clearly see if your line is balanced or not.

5. Check availability of WIP:

All operators must have enough work (WIP). You need to check WIP level at line level and operation level. Plan for future (next days) loading. If you know loading plan (style) you plan for other things. A WIP report will help you to track WIP status.

6. Prepare work schedule:

Though work scheduling is not a common task of an IE, you must have schedule about what all style to be loaded in the line in advance. You should thoroughly study the new style and analysis requirement.

7. Measure line efficiency and labor productivity:

To know the factory performance, measure and make line efficiency report daily. If possible, measure efficiency of the individual operators. By analyzing individual operator performance (efficiency %) you can find low performers. Compare daily efficiency with the target efficiency. Then plan accordingly to improve line efficiency. Measure labor productivity. Another measure of a line performance is labor productivity.

8. Motivate employee:

By measuring individual operator performance, you will know good performers. Make a list of good performer and praise for their good work. Prepare a list of monthly top performers and announce their name in front of the management. Other way to motivate shop floor employees is to provide them performance incentives. Nowadays technology is available for displaying individual employee's production quantity and efficiency.

9. Measure labor cost per piece

There is a target labor cost per garment for each style. This cost is the amount taken in garment costing sheet for direct labor cost. Your aim should be to beat that target cost and bring down per piece cost. That is why you need to calculate it daily and compare it with the target.

Practice above things and surely you can control the line output, production cost and line performance.

Why Measure Stitching Line Efficiency?

Line efficiency is one of the common measures analyzed by factories on daily. Line efficiency is one of the factory KPIs.

You might be measuring line efficiency and analyzing data on daily basis. Do you know why you should measure line efficiency?

Why measure line efficiency?

There are many reasons for calculating line efficiency. Some of those reasons are listed here.

- To compare actual line efficiency with the target efficiency of the factory
- To perform comparative study between stitching lines and floors (even product wise comparison) and motivate low performing lines.
- Production team and industrial engineers are responsible for maintaining line efficiency as well as average factory efficiency up to certain level. To know their line performance, they can't avoid measuring line efficiency.
- Actual direct labor cost is reverse proportional to average efficiency of a line. Labor cost can be reduced by increasing line efficiency. If factory measures its performance on daily or monthly, they can check how much they are improving their performance in each month.
- To analyze performance trend of the individual lines
- To find improvement potential and impact on saving by increasing line efficiency

Few other application of line efficiency data include

- Efficiency data is used in capacity planning for the factory
- Garment making cost estimation is done based on average factory efficiency.
- Calculating group performance incentive for stitching operators and other personnel involved in production.
- For developing learning curve of a line

I hope this list would encourage you measuring line efficiency and making efficiency report by production line.

How to Improve Finishing Room Efficiency?

In garment manufacturing, industrial engineering tools and methods are primarily used in stitching floor to improve labor productivity and line efficiency. A garment factory can implement the IE methods to improve finishing efficiency in the similar ways.

To bring improvement in the finishing room, you need to act on the following areas.

Calculate SAM: To measure actual efficiency of the finishing section, you need to establish standard minutes of finishing jobs (operations).

In the article *How to calculate SAM of a garment,* I have explained how to determine standard minutes (SAM) of an operation. The same steps can be followed for establishing SAM of the finishing process operations.

Calculate daily target of finishing tasks: Once you measure standard minutes of all finishing operations, you can do other calculation like daily finishing target, manpower requirement, and line balancing. All these will enhance finishing floor performance.

Conduct Time & motion study: Do motion analysis of the finishing activities. By doing this, you will get to know opportunity areas available in the finishing section for improvement.

Improve material movement: Brainstorm with your colleagues and think how you can reduce material movement time, waiting time, operation cycle time for finishing jobs. By reducing these you can improve finishing room efficiency.

Improve finishing room workstation layout: You can also think of improving workstation layout in the garment finishing section. By improving workstation layout, material handling time can be reduced.

Follow productivity improvement tips: In an earlier article, I have discussed 20 ways to improve productivity, various ways of increasing sewing operator efficiency. Follow the same steps for improving finishing room productivity.

Industrial Engineer's Digest

Selection of Improvement Projects and Project Implementation Method

Are you planning to do an improvement project in your factory or in departmental level by yourself?

I can suggest you many topics that can be taken as improvement projects. But just suggesting a topic does not make any sense. Because every factory is unique in term of problems faced by them. What all problems you are facing need to be solved one by one. Solving a problem following a proper method is a project. You should do an improvement project that is essential to improve your departmental performance as well as factory performance.

So, I would suggest you first learn how to select the right topic for mini projects (improvement projects) in your factory or within a department. Then go for implementation of that project with the predefined project plan.

I will show you guidelines on how you can select the right topic for improvement project and implement the same. Continuous growth is necessary to survive in today's business. Whatever department you are handling, you must keep doing new things that help to increase your performance. Bringing improvement in departmental level is important to keep yourself and your team motivated. Otherwise, if you only perform the same task day after day you will lose interest in your work.

A lot of things are happening around the world and in your neighbor factories. Change in management policies, implementation of new systems, new technology solutions, processes improvement to name a few.

One way of bringing improvement you can steal good practices from other companies and implement it to your company for the betterment.

Secondly, instead of stealing ideas from other factories there are a lot of things and opportunity areas that you can do to improve the performance of your factory and department that you head. Undergo a small project with your team. Once you learn the technique of problem finding and can design an action plan, it would not be a difficult job to do an improvement project.

Implementation of a project is done in three steps
1. Listing of real problems
2. Selection of the problems for improvement project

3. Project implementation

1. Listing of Real Problems

I have named it Real Problems. List down real problems those irritate you, those reduce your performance, and create delays, effect on quality. Consider problems as opportunity areas for you. There are two ways to find problems or areas that need to be improved.

- If you know there are problems within your department and those problems need to be sorted out, then list down all those problems. Each problem can be taken as project topics.
- If you do not find any issue within your department related to your performance but wants to do some improvement project, ask the head of the department (HOD) of other departments. Listen to what they think that you need to improve. List down what others suggest you improve.

2. Selection of topic

Now you have a detail list of problems in your hand. To select one topic from the list, follow these steps.
1. Open a poll with your team members and associated departments, ask them to list their topmost problems.
2. if you have measuring tools and have data for listed problems use those data for selecting your topic or
3. select a topic that you think is easy to accomplish but gives you better outcome.

Example#1: Take an example of Garment Production. If you ask me to find problems related to your factory performance. I will review Key Performance Indicators (KPIs) of the production department and analyze those KPIs to find which KPI need to be improved first. Common KPIs of a garment production factory are - Line Efficiency, Labor Cost per minute, Throughput time, Non-productive time, Productivity, Line set-up time and DHU level.

I am assuming that you measure your factory KPIs. Collect figures for each performance indicators for the last three months. Compare figures against your target or industry benchmark. Which KPI do you think need to be improved first? Or which one is easier to improve but might bring higher benefits to the company than other KPIs? Pick that one as project topic. Let's say average line efficiency is not as par your target (or at par industry benchmark). You can do a project on "Improving Line efficiency" and your problem is 'Low factory efficiency'

In case you do not measure factory KPI, then deciding KPI of your factory/production department and measuring each indicator itself is a big project. Do a project on this topic.

> Topic: **Implementation of KPIs and Analysis of Factory KPI**

Example#2: If you are working in the quality department then you can also do many things to enhance product quality. The most important thing is the factory must have a stable quality system and everyone follow the system religiously. If you already have a stable quality system in place, then analyze major quality issues reported by the buyer's inspector or internal quality managers. Pick one topic from buyer feedback and do a project on improvement.

> Topic: **Reduction of DHU of top 3 defects**

3. Project Implementation or Plan of Action

Project Topic: Improving Line Efficiency: Now the question is how to move forward to improve line efficiency? Project selection part is easier. The most difficult part is the implementation of a project. Sit with your team or you can do this part alone. When you focus on this project you need to find how line efficiency can be improved? What parameters are associated with line efficiency?

Read various methods that I have explained in the article **20 ways to improve productivity in garment production** and pick one method that suits you and go ahead with implementation.

Prepare an action plan once you have decided what methods you are going to do to improve line efficiency. Do implementation as per your plan. Assign the task to your subordinates and execute whether everything is progressing on time and getting targeted deliverable.

Measure line efficiency after improvement action has taken place to the line. Compare it with the previous efficiency level. If your improvement project helps to increase line efficiency establish it in your system and add changes to your routine. If your improvement project does not bring improvement in line efficiency, find another way to improving line efficiency and implement it.

Need Ideas for Improvement Projects – Check This List

Just doing daily activities is not enough. You need to do some extra work to improve the production process. For improving productivity or product quality.

I know, many industrial engineers used to do improvement projects one after another. You may be already implemented many improvement projects. Now you are looking for the next one.

You might be done a good project in the last year – but the same improvement project would not interest you and others for a long time – we want something new, interesting, and acknowledgeable work.

Do not stick with one project. Move ahead and become more creative and bring new ideas. To continue doing good projects we need inspiration. Some ideas may not be new to the garment industry - but it can be new to your factory. Or you can repeat a project work that has been done long back in your factory but currently not in practice.

Source of improvement project ideas

Here is a list for getting new ideas for improvement projects.

1. Talk to your friends who are working in the other garment companies as IE or in the other departments. Check with them what kind of development project they are doing. Keep discussing what is happening in the garment industry - their productivity benchmark and quality level. From the casual discussion with friends, you can find a fresh idea for doing an improvement project.

2. Attend seminar and workshops: If you get chance, participate in the seminars and workshops that discuss topics related to the garment industry, garment manufacturing, process improvement, system, technology, latest trends etc.

3. Find fresh ideas by reading trade journals. Many apparel trade journals are available online. You can read articles and news on fibre2fashion.com, textiletoday.com, apparelresources.com.

4. Read the published internship reports online. Students do project work during their summer/winter internship and publish their project report on digital libraries. In the students' internship report, they share their work, working method and project findings. You can refer to those for project topics and consider working on those if you find interesting ideas. Students' internship report may not be professionally

written - so what? You just need to skim the project report and need to pick the idea.

The best digital libraries and online platforms for finding published project work are -

- **Academia** - This is a web platform where students and faculty members publish internship report and papers.
- **SlideShare** – Presentations and project reports
- **Scribd** – A digital library. A place where you can find many internship reports uploaded by students.
- **onlineclothingstudy.com** – You can follow and read my articles on www.onlineclothingstudy.com and explore OCS to get new project ideas.

5. An improvement project can be like installing an automatic sewing machine in your factory for a specific operation. May be exploring the usability of the new machine. Developing new machine guides and attachments.

6. Follow industry trends. If you do not find time for reading project reports, trade journals – know about the latest buzz word in the apparel industry. Ten years ago, the buzzword for apparel manufacturers was LEAN manufacturing. These days we are talking about smart manufacturing, industry 4.0 manufacturing and circular fashion (sustainability in fashion manufacturing).

We all are innovative and creative by nature to some extent. But at some point, we face challenges in finding new ideas. How long one can be creative? You may think out of the box and get inspired by Google's projects or Elon Mask's project. But all big ideas cannot be implemented on the shop floor. Your project ideas must be relevant to your factory, to your business and product.

At the start, choose an improvement project that does not require much monetary investment. Show the improvement done through your project work. Later, you can pick projects that involve some expenses. You may need to do cost-benefit analysis of the project for fund approval. One idea from my side – you can work on the sustainability of your implemented projects.

Part – V
Advance Reading

Some Factories Do Things Differently

What the best apparel manufacturing companies do things differently?

In the recent past, I had a talk with one of my friends. He is working in a knitwear garment manufacturing company in Vietnam as the project head.

We were discussing different issues faced by industrial engineering department and production team in completing the orders on time. The common challenges of the apparel manufacturing business are lead time and manufacturing cost. And we were discussing the future possibilities of improvement in these areas.

I was inquiring about their shop floor performance and what they use to do for maintaining the benchmark performance. What are the expectations of their management from the factory IE team?

From the discussion, I came to know some very interesting facts that were a remarkable performance in the garment manufacturing industry. To be true, I have never heard of such kind of performance a garment unit may have. Especially, in the Indian garment factories, I have never seen or heard about such kind of performance and improvement they have done in the recent times.

I have listed some of their record-breaking performance in the following pointers.

Style changeover time is near to zero: Their production lines do not waste time in the style changeover process (line setup). Though they manufacture a staple product, their product includes a variety of designs and they need to setup the line for new styles. Relocation of manpower and machines are done. To maintain the style changeover time minimum, they train their operators in advance in the training area prior to assigning operators in the production line.

Most of the operators earn a lot of incentives (Money): They give performance incentive to their workers. Under the performance incentive scheme, their employees earn good amount of incentives. This is one of the primary reasons for being competitive. Their employees undergo a training program before working on the shop floor. In case an operator is given a new operation in which she has no prior skill, she just says 'no' to doing the new operation. The reason she would not be earning good incentives in the new operation. Operators demand skill training on that specific job (operation).

Factory efficiency 100% plus (on GSD standards): Often they reach above the 100% efficiency. The individual operator efficiency comes more than 100%. In that factory reduction of two seconds from an operation cycle time is counted as an improvement.

Cost saving projects

How do they make these possible? They have built a team to take care of the shop floor and improve the shop floor performance from all the possible ways. They have following teams.

Innovation team - This team works on a new and innovative way to improve methods and reduce the production cost in the coming years. Even two seconds reduction in the cycle time of an operation is considered and cost saving in the long run.

Kaizen team – They encourage employees to think about doing the work in a better way. Manage and evaluate Kaizen project. Organize Kaizen event and best performers are awarded.

Operator Training cell – Factory has an independent cell for training fresh and low skilled stitching operator. Even skilled operator get training on method improvement and doing a new kind of operations.

Conclusion

There is no limit of improvement in the garment shop floor. One should not be satisfied with 60% of line efficiency (monthly average efficiency).

Do you know what should be the benchmark performance of an apparel manufacturing company?

Do you have any goal on your factory performance, in terms of efficiency or production cost in each year? Are you happy with that?

My guess, you are not happy with the current factory performance level. Industrial engineers can do small improvement projects to bring small improvement in every day and in a year a huge improvement can be achieved.

I hope the points discussed in this article will inspire you in bringing such breakthrough improvement in your organization.

Which payment system is better- Piece Rate System or Salary based?

When an entrepreneur starts a business in garment manufacturing and garment exports, they look for a fair payment system that would motivate their workers, as well as factory can save good money out of the business. Whether payment system is salary based, piece rate or incentive system. Researchers and experts say that all payment systems have good things with certain challenges. After all, employers must understand what motivates their employee. Is it money, Job satisfaction, Position and/or recognition?

In this article, I will be discussing about sewing operators and assuming that they get motivated by money. When operators are paid according to the work done (pieces) by them with an agreed rate only then they will work with maximum effort. In another way, it can be said that the factory will only be able to produce a target quantity of garments when their employees are fully motivated.

Prior to choosing a pay system, the following factors must be considered. In the following table, a comparison of the piece rate and salaried system has been explained on eight parameters.

Parameters	Salaried	Piece rate
Labor rate per pieces	Lower	Higher up to 45%
Cost per piece depends on	Work content	Marketplace
Operator focused on	Quality	Quantity
Attrition rate	Lower	Higher
Capacity	Fixed	Varies with Labor availability
Product Change	No. problems	Price negotiation
Advance Technology	Welcome	Restricted
Work study	Welcome	Restricted

Source: Presentation by Methods Apparel at OGTC conference, 2011

Now, factory management needs to decide how they want to run their business. What their priority is?

Is it quality or quantity? Are they wanted to run the factory with fixed or varied capacity? Do they want stable operators, or it does not matter to them?

In the context of Indian exporters, they are still quantity conscious than quality, though they are losing a lot of money due to poor product quality every day. Indian exporters mostly prefer "Piece Rate System".

Benefits of "Piece rate" employees:
- Piece rate employees are highly motivated.
- Piece rate employees have no obligation about their employer.
- When factories have no work, they do not have to pay operators.

Challenges for having piece rate employees:
There are few challenges when a company hires "piece rate" employees.
- Design a fair rate for piece rate employees is most critical.
- Most of the factories do not have the system to establish a fair rate or standard time.
- Generally, the rate is decided through negotiation with employees.
- Until rate is fixed nobody work at their normal pace and the company loses capacity.

The salary-based systems also have many benefits and challenges.

Managing Individual Performance Bonus

Question: *Can you describe how bundle tickets should be created so that individual performance bonus system can be implemented. What operations should be grouped together for what bundles? ...asked by OCS reader.*

Performance-based bonus system

In a performance-based bonus system, operators become eligible for bonus earning when they reach the defined target performance level. This means producing garments more than the target set for bonus eligibility criteria. The performance of sewing operators normally presented in efficiency percentage.

For individual operator performance bonus, you need to track production (in pieces) of individual operators and calculate SAM produced by individual operators in the given worked hours. You can manually collect daily production of each operator. Operators may work a single operation or multiple operations. Later, multiply production quantity by operation SAM to calculate produced SAM by an operator.

Let's say, a factory sets up bonus earning criteria as this – when an operator reaches 60% individual efficiency, operator will earn a lumpsum bonused of ₹100. And after 60% efficiency, for each point of efficiency increase, operator will earn ₹2.00 as performance bonus. If an operator performed at 75% efficiency, her bonus earning will be sum of jump bonus plus bonus earning for extra efficiency. Her extra efficiency is 15%.

Bonus earning = (100+(2 x 15)) = ₹130.00

Bundle Ticketing

Bundle tickets are paper tags attached with bundles. Each bundle ticket contains bundle information like style number, layer number, bundle number, layer sequence, color, size, and garment quantity in the bundle.

In the garment manufacturing, RFID-based and Barcode based real-time systems are available for shop floor production tracking. In both systems, bundle card and barcode stickers are attached to the bundle as bundle ticket. In some of the RFID based systems, once the RFID tag is being scanned in the workstation (card reader) individual efficiency figures as well as the bonus amount is automatically calculated by the system and displayed on operator consoles (terminal). The bonus amount is calculated based on your input (bonus scheme) and operator efficiency on the day.

Similarly, in barcode tickets, the system generates individual operator efficiency once barcodes are scanned by a scanned and data is sent to the system.

Sometimes factories face issues in capturing correct production quantity done by individual operations where more than one operator work on the same bundle. If you follow bundle ticket, you can't get correct production of each individuals. This may cause in incorrect employee efficiency and in turn affect operators bonus earning.

The second cases found in the assembly line when operators do not follow bundle system. In this production flow, each operator completed one garment and pass it to the next operator. In such situation, when more than one operator assigned to the same operation, it would be difficult to count actual production quantity of those operation. In such situation, single piece flow (single piece bundle ticket) can help you getting operators' correct performance. Or make small bundle size (5 units per bundle) and ask operators to follow bundle from start to end. Operators should not split the bundle.

You need to ensure that operator scan bundle tickets or stick barcode tickets on the gum sheet after completion of a bundle.

Grouping of operations

Grouping of operations is done during line setting to make all operations with a limited number of operators. For example, if a style has 45 operations and you want to set-up the line in 30 operators then some operations need to be clubbed. You can keep all 45 operations in your operation bulletin and have different SAM for each operation. When one operator does two operations, count production operation wise. SAM of these two operation be different. Calculate operation-wise total same produced and then sum up operation wise produced SAMs to get total produced SAM by the operator.

When you need to group operations, first consider machine types and ensure that both operations can be done using same sewing machine. In such case, an operator can perform both the operations in the same workstation.

There is another way of grouping operations. Assign two consecutive operations to the same operator. In case the given operations required different machine types, assign two machines to this employee at the same workstation. Operator will complete one bundle in first machines and then rotate and sew the second operation in the second machine.

Secondly, for line balancing, you may need to club operations or separate multiple operations. To reduce issuing of multiple barcode sticker or tags you can convert a group of operations into a single operation.

For calculating bonus earning amount accurately, employee wise production tracking should be done correctly. Operator can write their production quantity in a notebook for ensuring that they are getting correct bonus amount.

Line Efficiency: A Magical Measure for Garment Manufacturers

Line Efficiency. Also called as sewing line efficiency.

How do you use efficiency figures in your factory? Is it only a factory performance measure? Is it the key word for promoting your company's performance?

Do you feel that you are not sure how to deal with the factory's line efficiency?

I assume you measure line efficiency in your factory. What do you do with your line efficiency data? Is it just to compare one line to another in a floor producing the same product? Is it for per minute labor cost set-up? Is it for the capacity planning of your unit?

To me, efficiency is the Mantra, a magical measure (performance indicator) for a garment factory. Through continuous line efficiency improvement, one can build up his business and make a massive profit. One can bring discipline to the organization. With this single measure, one can control the cost of the company. I will show you the reason why it is a Mantra to me.

Should I repeat the formula and the method how to measure line efficiency? I love to repeat it.

> Line Efficiency % = (Total minutes produced by a line / Total minutes worked by operators in that line) x 100

Managers who are indifferent with the line efficiency or factory efficiency data, does not know his actual potential to make money from the manufacturing business. Few managers are in debate if their calculated efficiency is correct, and it represents the true performance of the factory.

Do you know how many total minutes you used to make an order? It is directly linked with your direct labor cost. If you do not know these figures or do not measure it, then you do not know whether your labor payroll is correct or not.

What is the minute value of an order?

Study the difference between two measures – the minute value of an order and total minutes consumed to make the order. You will find there is a big gap. This gap is the treasure for potential improvement. Do you know why this gap is there? This gap can be controlled and reduced by controlling efficiency figures.

If you know the magic of "Efficiency" figures, you will not think much on implementing performance-based incentive for the workers. The efficiency of a line can be improved further by operators' effort and managerial skills of the supervisors. Few good things about performance-based incentives.

- An appropriate incentive means happy employees.
- Happy employees make managers (owners) happy by producing extra garments.
- Increase in efficiency means a reduction in cost per garment.
- Not only the labor cost, but the whole factory cost per minute will come down through increasing efficiency level of sewing lines. Cost per minute of the factory is calculated as Total Payroll and Expenses / SAM produced in a defined duration. Increase in line efficiency means increased produced SAM in a period. Hence, as you increase the total produced SAM in a month, factory cost per minute gets reduced. It means lower in cost per unit. It results in a higher profit margin for the company.

Fact – the maximum reach of efficiency depends on the order quantity and style run (number of production days). So, do not assume all style have the same manufacturing cost per minute. Here you need to design quantity wise average efficiency rates for your factory for the cost reference. This efficiency database can be used in costing garment correctly.

Another KPI of an organization is productivity per labor per day. Labor Productivity increases when line efficiency is improved.

Some factory managers do not bother to check the efficiency benchmark. They think they perform better than the other companies do. As they are making enough money with the existing system, they do not bother to point out efficiency data. These managers do not find enough money to share with employees and increase employee salary yearly.

In summary, the major benefits of controlling efficiency include

- Reduction of manufacturing cost
- Accurate product costing based on the order quantity

- Employee motivation is possible through sharing profits earned through efficiency increment. Happy employees, less labor absenteeism, and improved employee retention rate.
- Improved factory capacity. As a result, more option for revenue generation with the same capital resources.

You see I call the efficiency measure a **Magical Measure** for a garment factory. I hope you can do the magic in improving factory's efficiency with the guides you learnt from this book.

Work Study and Industrial Engineering Terms and Definitions

In a garment factory, work study officers and industrial engineers use various technical terms like Basic time, Standard time, predetermined time, relaxation allowances etc. But due to unavailability of the reference books, they find it difficult to explain these terms correctly when required.

Some of the important terms and definitions related to work study department are provided here for your quick reference. Source of these terms and definition – *"Introduction to Work Study, Edited by George Kanawaty, ILO, Geneva, 1992"*

Work Study

Work study is the systematic examination of the methods of carrying on activities so as to improve the effective use of resources and to set up standards of performance for the activities being carried out.

Method Study

Method study is the systematic recording and critical examination of ways of doing things in order to make improvements.

Work Measurement

Work measurement is the application of techniques designed to establish the time for a qualified worker to carry out a task at a defined rate of working.

Work Sampling

Work sampling is a method of finding the percentage occurrence of a certain activity by statistical sampling and random observations.

Work Content

The work content of a job or operation is defined as: Basic time + relaxation allowance + any allowance for additional work – e.g. that part of contingency allowance which represents work.

Time Study

Time study is a work measurement technique for recording the times of performing a certain specific job or its elements carried out under specified conditions, and for analyzing the data so as to obtain the time necessary for an operator to carry it out at a defined rate of performance.

Qualified Worker

A qualified worker is one who has acquired the skill, knowledge and other attributes to carry out the work in hand to satisfactory standards of quantity, quality and safety.

Element

An element is a distinct part of a specified job selected for convenience of observation, measurement and analysis.

Work Cycle

A work cycle is a sequence of elements which are required to perform a job or yield a unit of production. The sequence may sometimes include occasional elements.

Rating

Rating is the assessment of the worker's rate of working relative to the observer's concept of the rate corresponding to standard pace.

Standard Performance

Standard performance is the rate of output which qualified workers will naturally achieve without over-exertion as an average over the working day or shift, provided that, they know and adhere to the specified method and provided that they are motivated to apply themselves to their work.

This performance is denoted as 100 on the standard rating and performance scales.

Basic Time

Basic time is the time for carrying out an element of work at standard rating, i.e.

(Observed time x Observed rating)/ Standard rating

Selected Time

The selected time is the time chosen as being representative of a group of times for an element or group of elements. These times may be either observed or basic and should be denoted as selected observed or selected basic time.

Relaxation Allowance

Relaxation allowance is an addition to the basic time intended to provide the worker with the opportunity to recover from the physiological and psychological effects of carrying out specified work under specified conditions and to allow attention to personal needs. The amount of allowance will depend on the nature of the job.

Standard Time

Standard time is the total time in which a job should be completed at standard performance.

Predetermined Time

A predetermined time standard is a work measurement technique whereby times established for basic human motions (classified according to the nature of the motion and the conditions under which it is made) are used to build up the time for a job at a defined level of performance.

Work Specifications

A work specification is a document setting out the details of an operation or a job, how it is to be performed, the layout of the workplace, particulars of machines, tools and appliances to be used, and the duties and responsibilities of the worker. The standard time or allowed time assigned to the job is normally included.

7 Must Read Books for Industrial Engineers

One can learn new things, enhance his/her skills, and become smart in doing a job –
- by undergoing a professional course and undergoing training
- by reading subject books and applying the learning to work
- by working on the job under expert guidance

Out of these, you have full freedom to learn when you choose to read books. Books listed here will help you in improving your knowledge while working as an industrial engineer (work study officer) in the ready-made garment industry.

The listed books are generic and may not be specific to the garment industry or industrial engineering. Theories you learn from reading these books can be applied in the garment industry. Like process improvement, method improvement, managing team and work, doing better planning etc. Anyone can read these books and enhance their knowledge and apply the knowledge in garment manufacturing. Here is the list for your reading and honing your skill. Most of these books are available on Amazon store for purchasing online.

1. Introduction to Work Study by ILO, Editor: George Kanawaty

This book was first published in 1957. This book describes the basic techniques of work-study as practiced in many parts of the world, has been widely recognized as the best available introduction to the subject for work study practitioners, teachers, and students. It covers method study and work measurement procedures and covers not only machine shops but also process industries, the services sector and office work.

2. Time and Motion Study by Ralph M. Barnes

You know time study and understanding motion economy is must for a good industrial engineer. In apparel manufacturing course, these topics are covered superficially. But you must have in-depth knowledge on the topics covered in this book.

3. The Goal: A process of ongoing improvement by Eliyahu M. Goldratt

This book is a nonfiction novel. Authors of this book explained process management of manufacturing industry by a story. This is a brilliant book that teaches process improvement. Authors teach process management, managing bottlenecks in the production and WIP with real-life studies.

4. The Toyota Way by Jeffrey K. Liker

From the book cover, "This book will give you an understanding of what has made Toyota successful and some practical ideas that you can use to develop your own approach to business."-- Gary Convis, Managing Office of Toyota. This book covers the fundamentals of Lean manufacturing. This book explains lean as thinking rather than tools.

5. Apparel Manufacturing: Sewn Product Analysis by Grace I. Kunz, Ruth E. Glock

This is a must-read book as you also need to learn about garment manufacturing processes and technology, with the industrial engineering concepts. I read this book in my college days and have one copy in my library. I learned a lot about garment manufacturing technology reading this book.

6. One Minute Manager by Kenneth Blanchard PhD and Spencer Johnson M.D.

This book is an old one and written in the 80's. But it's one of the best books ever written on how to manage people and performance. By reading this small book, you will learn how to manage your subordinates (team). Why this is important to take time explaining the work first and instead of just telling your people what to do? The information in the book can be practically used by all managers.

7. Industrial Engineering in Apparel Manufacturing: Practitioner's Handbook (The Book) by Prabir Jana and Manoj Tiwari

I loved the content of this book and in-depth theory coverage. In this book, authors have covered all the essential industrial engineering subjects (related to apparel manufacturing) that we practice on the shop floor.

Apparel industrial engineers, students and production managers who want to learn more and enhance their knowledge on industrial engineering and production management can read this book. Garment factory owners can gift this book to their IE team.

365

Industrial Engineer's Digest

Sewing Machine Ratio for a Shirt Factory

Question: *I am in the process of setting up a garment manufacturing unit in Bangalore. I will begin by doing Job Work for some big players. I want clarity regarding the machine ratio. I have been offered orders for Men's Shirts. I want to begin with around 30 machines then gradually increase it. I want your valuable advice on the ratio of different types of machines I need to buy. Kindly advise... asked by an OCS reader.*

It is not enough just to know different types of sewing machines required for making shirts. You need to know machine type, how many machines you need to buy for each type of machines. When you are going to set up a shirt factory, selection of correct machine ratio is important to maximize utilization of your machines and capital expenses.

Here, I am explaining how machine ratio is calculated.

Machine ratio is normally calculated from daily production target (quantity) and product styling.

Where, production target is linked with following factors

- Line Efficiency
- Number of workers (Sewing operators + helpers)
- Product SAM
- Daily working hours
- Order quantity
- Product difficulty level

You have already decided total number of machines that you would purchase i.e., 30 machines. I will show machine ratio for total 30 machines.

In another article, *How to Determine Machines Requirement for a New Factory*, I have shown how one can calculate machine requirement from product with example of four basic product. (Shirt, Trouser, Tee and Polo).

Here, I will show you how to calculate machine requirement for daily fixed production target. Calculate Shirt SAM and calculate daily production target based of shirt SAM and available machines. Let us say garment SAM is 20 minutes and average line efficiency is 70%. According to these information, your daily target for shirt production will be 504 pieces.

Daily production Target = ((Number of machines x 480)/Garment SAM) x Line Efficiency%

Daily production target = ((30 x 480)/20) x 70% = 504 Pieces

Now, in the operation bulletin, calculate machines requirement for making the daily production quantity (504) for all operations. I am not showing the operation wise machine requirement breakup. Read another article for details for calculating operation-wise machine requirement.

Based on the calculation, you will need the following machines. 6 different types of machines and number of machine in each type is shown in the following table.

Table: Machine ratio for making shirts (30 machines in total)

Sl. No.	Machine Types	Quantity
1	Single Needle Lock Stitch (Plain with/ without UBT)	24
2	Single Needle Lock Stitch machine with Edge cutter	2
3	Multi-needle Chain Stitch Machine / Kansai (Optional)	1
4	Feed-off-the-Arm (FOA) or 5 Thread overlock machine	1
5	Button stitching	1
6	Buttonholing m/c	1
	Total	30

In case you want to add other type of sewing machines (i.e., Special machines) then consider replacing that machine with single needle lock stitch machine or respective operation to keep total number of machine 30.

Relationship Between Line Efficiency and Labor Costing

Industrial engineers calculate direct labor cost (sewing labors) of making a garment. Labor cost per unit directly related on the line efficiency (performance). When line efficiency increases, production volume of the line increases. So, the cost per garment goes down when line efficiency increases.

Normally, industrial engineers calculate per minute labor cost at 100 percent line efficiency and based on operators' monthly salary.

Labor cost per minute increased as line efficiency goes down. Many engineers make a matrix of line efficiency Vs. labor cost per minute (See the following table for an example of such matrix).

Once they have labor cost per minute against line efficiency as shown in the following table, minute cost is multiplied by garment SAM to calculate garment cost. Using the simple formula, you can also calculate labor cost per minute of your factory.

Table: Labor cost per minute in $ (An illustration)

Monthly operator's salary	$200		
Total minutes in a month (26 days x 8 hours shift)	12,480 Minutes		
Efficiency%	Labor cost/Minute	Efficiency%	Labor cost/Minute
100%	$0.016	45%	$0.036
80%	$0.020	40%	$0.040
65%	$0.025	30%	$0.053
50%	$0.032	20%	$0.080

Assume that you have a line of 30 operators, and all are salaried workers. So, you pay fixed amount to your workers instead of their performance level. At present line perform at 30% and daily output is 300 pieces and if daily salary of the operator is USD 5.0, per piece labor cost becomes (30x5/300) = $0.50 per unit.

Suppose efficiency increased up to 60%, at this efficiency level line will make double garments i.e., 600 pieces per day. In this case the operator's daily salary remains the same.

So, per garment labor cost now become = (30x$5/600) =$0.25

Here, I have shown you straight method. Average line efficiency of a product varies depending on the order quantity per style and product construction difficulty level. So, the cost of the garment varies accordingly.

How had I Learned Industrial Engineering (and How You Can Learn Too)?

For the last 9 years, I am writing tutorials on industrial engineering topics ... and I learned many things on industrial engineering through this process. Once an OCS fan asked me this question - *how did I learn that much of industrial engineering subject knowledge?*

I will share methods that I used to learn industrial engineering and still using. These are non-traditional learning methods but very effective. You can copy my techniques and learn industrial engineering.

Learning IE by doing:

I do not have an industrial engineering degree. Many things I know and learned about industrial engineering is by practicing it in a factory. I read industrial engineering topics in college. I had seen many IEs to practice industrial engineering methods and use IE tools in factories during my internship and graduation project. When I was heading the production planning department in an export house, I used to implement basic industrial engineering tools in the factory. I did many mistakes initially, but I learned from my mistakes.

So, you can too learn industrial engineering by doing and practicing it on the floor. While doing something first time, you might make mistakes. That is the way to learn things. Keep it up. I learned production planning and merchandising also by doing it.

Learning by teaching others:

I love to teach learners. Earlier I used to teach my subordinates. When I was a consultant, I taught many participants from client companies. When I use to teach others, I learn many things in the teaching process. You might have read IE tutorials (articles) on OCS blog. Do you know how I learned all those topics? All because of OCS readers who force me to learn, so that I can teach them later and can write new tutorials. I get questions on various topics from OCS readers. I prepare the answers and when I prepare answer for a specific question, I need to research and read the topic first. Thus, I learn it.

To learn IE, you can start teaching your subordinates or juniors. You do not need to be master in your subject to teach others. Whatever you know more than others, you can start teaching with that only. When you start teaching, you will be asked questions by your students. It is not necessary that you need to know all answers. If get chance, you can learn the answer to the question and answer to your students. If do not find your student start teaching others from production department.

Learning by asking questions:

This is the easiest way of learning something. When I met with industrial engineers in garment export houses, I ask different types of questions to them for clarification. I get answers for most of the questions, as they are working on this field for many years, they know the answer. You can also start asking question to your seniors when you need to know something. If you do not have a senior or experienced colleague, you are welcome to send your questions to me.

Reading articles online and/or offline:

I read articles on magazines and on the Internet daily. I learn many things by reading. I discover new areas of learning by reading. My reading topics are not limited to apparel production and industrial engineering only. I read various topics to get ideas.

You can also start reading technical magazines related to apparel production management and industrial engineering. If you do not like to spend money on print magazines, read free articles on the Internet. By reading articles on industrial engineering and apparel production management, you can increase your knowledge and you can teach others a new topic.

Discussing topics with your colleagues and seniors:

This is a great way of learning. Currently, I do not have industrial engineering colleague. But I spent two and half years with a group of IEs. That time we had a lot of discussion on industrial engineering. Still, I practice this process of learning. I said earlier, I regularly use to visit garment export houses. When I meet with factory IEs, we discuss things whatever we find interesting, within IE topics or out of industrial engineering topics.

Watching videos on YouTube:

YouTube is a great source for learning by watching videos. You will find numerous videos on sewing operations for different products, semi-automatic and fully automatic sewing machines, and different types of material handling equipment. When you watch videos, observe it closely. You can learn many things like, good movements, workstation layout, working speed, machine types, attachment and guides used and many things.

You can even get knowledge on latest innovation in sewing machines and fashion manufacturing field.

Conclusion:

There is no end of learning. And nobody can improve his/her performance continuously without learning. Here, I have explained how to learn industrial engineering subject. But these techniques are applicable to anything you like to learn. These days there are online courses available for learning any subject that you like to learn.

Difference Between Basic Time and Standard Time

Basic Time and **Standard Time** are used in work measurement. Observed Time and Cycle Time are two other terms used in time measurement. To differentiate these two terminologies, definition and explanation are given below.

Cycle Time:

Cycle time is defined as the time duration from the starting point of the operation cycle to the starting point of the next operation cycle. This time is established from the observed time records while doing cycle timing an operator while working at a certain pace.

Example: In a sewing operation, from the moment of first piece picking-up to the picking-up the second piece, is called cycle time. In the cycle time, material handling time is included with the actual work time in a machine.

> Cycle time = (Machine Time + Material handling time)

Cycle time is also known as Observed Time.

Basic Time:

Basic time of a job is determined by multiplying rating factor to the observed time (cycle time). Basic time is also called as Normal time.

> Basic time = (Observed time x Performance Rating)/100

In Basic Time, allowances are not included.

Standard Time:

Standard time is the time allowed to an operator to carry out the specified task under the specified condition and defined level of performance. This is a standard definition for standard time.

Some additional time is added to the basic time to arrive standard time of a task. In practice, none can work throughout the day without taking rest. Operators need time for relaxation from fatigue. Various allowances are relaxation allowance, contingency allowance (like machine breakdown) and bundle allowance (for PBS system).

> Standard Time = Basic Time + Allowances

The basic constituents of standard time are shown in the following chart. This chart shows how standard time is made up from the observed time and basic time of a job.

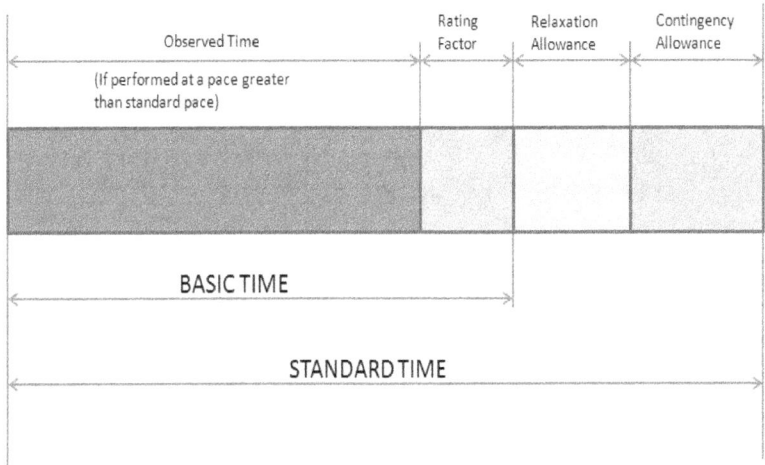

Chart: This chart shows how standard time is made up

For a specific task following conclusion can be drawn
- Observed Time may be greater or lower than Basic Time.
- Basic Time is always lower than Standard Time
- To set Standard Time of a job, you must add allowances to Basic Time (normal time)

Pros and Cons of Setting up Garment SAM Higher than the Standard Time

In the previous article, I have discussed estimating garment SAM based on IE's experience. In the experience-based SAM calculation method, you know that there is a possibility of establishing a low/high standard time (SAM) of a garment than the actual standard time.

But what happens when a factory intentionally sets up garment SAM higher than Standard time? After reading this article you will know the Pros and Cons of setting up garment SAM higher than the standard time.

You might be thinking why should one use to show increased SAM? If you do not aware of this fact, this is for your information that such practice exists in the garment industry. In one of my factory visit, I had questioned the IE manager of the company on the SAM which seems higher for the specific product. IE Manager disclosed that the SAM figure on the OB sheet is set higher for buyers.

Pros of Higher Garment SAM

There is only one benefit of setting up garment SAM higher than actual SAM. And that is quoting a higher labor cost in the costing sheet. Now, many apparel buyers are aware that labor requirement for making a garment is measured by SAM. There is a direct relation between garment SAM and direct labor cost.

So, even if buyers ask for open garment costing, and when buyers negotiate for FOB, garment factory can justify their production cost.

Cons Higher Garment SAM

On the other side, there are many disadvantages for setting up garment SAM higher.

1. Suppose that the factory considers the increased SAM for calculating factory performance and other SAM based calculation. The outcome will be as following -
    - Factory will get a totally wrong data of the line efficiency and individual operator efficiency.
    - The calculated salary for operators will be higher than the actual salary (if calculated based on earned SAH).

- Performance-based incentives are calculated based on produce SAH. With higher garment SAH, operators will be producing higher SAH.
- All calculation, those are derived from garment standard minute, will be affected. Like, production capacity calculation, production scheduling.

2. To measure actual performance and getting the right calculated results, the factory needs to maintain two sets of records – one for the buyer and another for internal usage. An adjustment can be done for calculations, but such practice would complicate the IE's activity.

3. Standard time is used for improving the working method. Standard time (SAM) is used to compare the existing performance of the operators and standard time. If the Standard Time is not correct (as said it is intentionally set up higher), you cannot measure operator performance correctly.

4. Also when you prepare initial line balancing and line layout, you are setting the production target at a lower level. In turn, the calculated manpower and machine requirement will be higher than the actual requirement.

Industrial Engineer's Digest

Key Qualities Needed to Be a Successful IE Manager

Question: *What are the key qualities one needs to be a successful IE manager in the manufacturing industry? ... Asked by an OCS reader*

You asked about key qualities to become a successful IE manager. If I am not wrong, you are already an Industrial Engineer. If you are an industrial engineer, you must have the desire to become an IE manager in the coming days. If you have spent sometimes in working as an IE, you must have acquired skills and abilities to become an IE manager in your field. Still, there are some essential qualities including managerial skills and interpersonal skills which are important to become a successful manager inspective of the field of areas. All these skills can be learned.

One must have the following qualities to become a successful IE manager. I will take an example of the garment manufacturing industry to explain qualities of an IE manager.

- Good at mathematics and numbers. Every day he needs to play with numbers, like target, efficiency, productivity, overtime, bonus, costing, SMV, learning curve, planned efficiency, line balancing, etc.
- Must have a learning attitude and accept the changes.
- Understand product construction (garment construction) and production specifications well and have clear knowledge about the complete manufacturing process flow.
- Good at data analysis and good at working on Excel sheet
- Must have knowledge and know-know of industrial engineering/work-study tools, machines, and equipment used for making garments.
- Should believe and practice on tools like – The checklist, SOPs, To-do list, Scheduling and keeping Journal.
- Ability to motivate subordinates and shop floor employees
- Ability to train others and good at giving instruction to his subordinates and workers.
- Must strive for the best and do continuous improvement.
- Have time management skills, pro-active and good at planning and execution of a task.

- Good interpersonal skills, communication skills, work presentation skills

Other than the above qualities, check the list of required organizational skillsets for becoming a manager. You can learn those by reading couple of management books.

Manoj Tiwari, who is an IE practitioner and an academician preparing students for becoming better industrial engineers, recommends a few additional qualities (for sustainable improvements) for the IE managers.

- IE must be practical in approach. He should be good at how the improvements can be carried on the floor. There is not always 2+2=4 in practical environments. Of course, and IE need to have the mathematical and logical aptitude but also must be aware of ground realities.
- Whatever, improvements (maybe in terms of setting up targets, method improvement, process improvement or anything else), and an IE should be capable of making it happen on the floor. This adds to the credibility and acceptance of IE and improvements shall be sustainable.
- Further, many of the IEs only focus on sewing. In my opinion, an IE should focus on overall process improvement, as there is huge potential for improvements (with IE interventions) in other areas as well such as Fabric & Trims store, Spreading & Cutting, Dry Process, Washing and Finishing & Packing as well.
- And last but not the least and IE should be Tech-savvy, always ready to adapt newer/advanced tools, scientific ways, and next (best) practices, as that save efforts, time, and resources immensely.

I hope the above guides will help newcomers and young engineers to prepare themselves in becoming a manager in a garment manufacturing factory.

Identify the Reasons for Low Line Efficiency

On the shop floor, every day the production and IE team work hard to improve the line efficiency. At least they try to maintain the monthly average line efficiency (benchmark efficiency). But the efficiency level goes down from time to time. In this article, we will try to identify the reasons for low efficiency of the factory.

What is the average efficiency of your production lines? Let us assume the average line efficiency of your factory is 60% (Monthly average). If this is true, how is the balance 40% standard time lost? Where that 40% of the standard time is gone? (Note: actual line efficiency data might be different to your factory).

You need to study the fact, analyze data, and find out the root causes for having low line efficiency.

Ask these questions to yourself and note down the answers to identify the reasons how a factory losses its line efficiency.

	Reasons (root causes)	Questions to ask	Yours Answer (Y/N)
1	A dry line	Do you often find that most of the operators in the line sit idle due to a shortage of feeding?	
2	Idle time	Do you see operators sit idle for some reasons? If so, what are the reasons?	
3	Stitching quality issues	Do you need to stop operators if they are not making desired quality? Do you send the defective garments back to the line for repair work? Does it build a bottleneck in the line? Think of an alternative way of repair work?	
4	Line set-up efficiency loss	Does your line take too much time to set up it for a new style?	
5	Balancing loss /An imbalanced line	Do you problem is line balancing? How much time do you loss due to balancing loss in the OB?	
6	Operator absenteeism	Do you often see some of your operators are absent? Do you have enough floaters in the place of the absentees? What you do to reduce the absenteeism rate?	
7	Off-standard work	Do you need to move operators from their job to another job on which they do not have the skill?	

8	Low employee motivation	Money is one of the best motivators to motivate workers and employees and it works. Do you have a performance incentive scheme for the shop floor workers?	
9	Overtime work	It is a common practice that many of us work 2-3 hours after regular shifts. Does the excess overtime reduce operator's performance? Think – can you stop working overtime by producing daily target qty within the regular shift hours?	
10	Low skilled operators	How many of the employees do not have the required skill? Do you have any plan to train them?	
11	Unmeasured time	Do you have the record of all lost-time hours? How much time is not counted in line efficiency calculation?	

You may have a good IE and production team who can identify the root causes of low efficiency of the line.

Identify the reasons and prepare a plan to work on improving the average efficiency level. In an earlier article, I have covered how to improve things on these points. You can follow those steps to improve efficiency of your factory.

Characteristics of a Good Workstation Layout

Prior to doing the line layout of an apparel item, you may need to design workstation layout for each individual workstations in a line.

The design of the workstation layout widely varies from one operation to another depending on the size of work (cut panels), the number of components to be worked on and type of machine to handle. However, few things are common for each workstation those are listed below. An example of a good workstation layout has been shown in the following image.

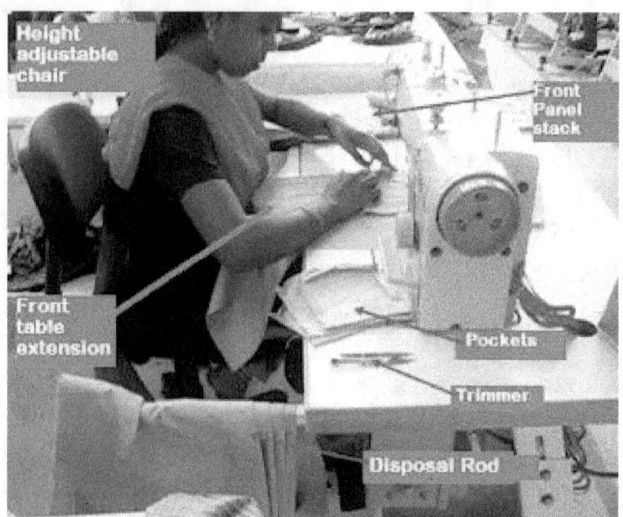

Workstation layout: Attaching front chest pocket in a shirt panel

Characteristics of a good workstation layout:

- All works (components) should be placed as closer as possible to the operator for easy reach.
- Work should be presented correct way for easy unbundling and quick pick up
- Nothing should obstruct an operator during picking up and disposing stitched components.
- Keep bins or other means to dispose the finished work.

- Operators must have enough space on the table to handle the garment parts. If required add table extension near operator and left side of the machine.
- Enough space in between two machine for operator movement.
- Use jig and folders where required.
- If you provide trimmer (for thread cutting) to operator define a space for it.
- Provide height adjustable chair to the sewing operators.
- The centre table (feeding table) must be placed in the left side of the sewing operator for easy picking-up of the work.

So, when you will be preparing workstation layout for a sewing operation, consider above parameters. With these you also need to consider the motion economy and ergonomics for preparing a good workstation layout.

Operator Training and Deployment in Apparel Industry - A Systemic View

According to Paul Collyer, if correctly deployed and resourced, trainers will reduce training times for recruits and will also improve the performance levels of existing low-performing workers both in terms of output and quality. As per my request, Paul has written this article.

In recent years, many companies in India and the South-East Asia region have started to use 'trained trainers' and have benefited as a result. However, observations show that the majority, although achieving substantial cost improvements are still "missing a trick" and not receiving the maximum benefits possible by not integrating training completely into the production management process and modifying their training systems to their needs.

This situation is due to a combination of factors, but a key point is the inability to appreciate how to modify the systematic training concepts to different production systems and circumstances.

The apparel industry is very diverse with many different products, production systems, order sizes and cultures to cope with. As with all aspects of running a successful garment business, there is no such thing as a "one size fits all" solution.

Those companies that have successfully embedded trainers into their factories should now in the pursuit of continual improvement look at moving up to another higher level by tailoring the training approach even more closely to their individual needs and circumstances.

Initially, if we first look at the training of freshers or new recruits.

- Are recruiting levels linked to both production requirements and initial training capacity?
- Are those two issues identical to avoid a shortage of recruits or the reverse; trainees sitting around or being used as helpers for several weeks resulting in losses?
- Does HR have a recruitment target that is linked to the previous two requirements?

The systematic operator training approach works under the following philosophy:

1. You train someone because you want them to perform a task(s) for you.

Therefore, you only train them in the skills and knowledge necessary to perform that task. Any additional training is both wasteful and counterproductive. The production system is paramount when designing the training program.

- Does the operator need to know one or more operations?
- Do they need to change style frequently or are they making a standard product?
- The approach need is to be modified for circumstances.

2. Sewing and handling skills are taught to the trainee; not operations.

If the trainee acquires the correct skills then they will be able to perform the tasks or operations. This is a key concept that determines the ongoing training program. If trainees are going to work with "fashion" items such as dresses and potentially on short production runs, then it will not be possible to design training exercises for every potential design feature and eventuality. Accordingly, a program to teach "core" skills will be required. Alternatively, if a standard product such as men's formal shirts is to be produced then a limited number of "core" skills allied to operation specific exercises is possible.

In addition to the product, the manufacturing system must be considered when designing operator training program. If PBS lines are in use then it possible that only one or possibly two operations must be mastered thereby reducing the number of skills to be learned, only teach what is necessary!

A team or modular system will require operators to learn many skills and operations and a step by step carefully planned approach needs to be taken.

Where "single piece" production teams are in use then again, a minimum number of skills will need to be learned. All systematic operator training is based on competence. Based on the ability to perform a task to a set standard. With sewing, this relates to output and quality and it is customary for trainers to have times for training recruits and then output targets as they move to production lines. In a PBS system management can set a minimum acceptable performance level and build these into their production balancing.

As "single piece" requires adjacent operators to work at different performances to meet the target the trainer can set individual competence levels in terms of time and speed to that required. i.e. trainees being prepared for different operations will have different target performances.

All the above demonstrates that the concept of systematic training needs to be modified for the circumstances within a factory and the classic systematic approach as shown below is effective but needs to be tailored for needs.

Systematic Operator training will be well known to those companies currently using it and should utilize the following program:

- Induction both HR for general employment issues and the trainer for specific production matters.
- Machine knowledge
- Foundation skills training. (speed and rhythm)
- Operation specific skills using a series of purpose, designed fabric exercises. It is at this point that customization for specific circumstances needs to take place.
- Production work, stock items or mockups per company policy.

Operation specific skills

As previously stated, these are taught using a series of fabric, (never paper) exercises. Each exercise teaches several skills and builds upon the learning of the previous exercise.

Each operation should be analyzed and the necessary handling and sewing skills identified. All operations will contain several "core skills" such as align at the foot, sew a straight seam, or stop accurately in a corner. Accordingly, several exercises can be developed to teach these skills and the appropriate ones taught to the trainee. Again, do not teach skills that are not required. If a recruit is to sew a collar, there is little point in teaching long seam sewing skills.

Once the "core" skills have been mastered then operation specific exercises can be given to allow the trainee to relate the skills learn to the task to be performed and to allow the practice.

The above approach is effective when standard products such as 5 pocket jeans or tee shirts are to be produced or if long production runs are envisaged and an operator will be required to perform one or two operations only. (Never teach two operations simultaneously; teach consequentially remembering that some "core" skills may be common to both operations.)

However, a different approach must be taken when making fashion items and the operator needs to be versatile. Products should be analyzed for skills required and "core" skills taught using the number of exercises. Again, teach only one skill at a time but remember that exercises will transfer core skills the next one to another and should if correctly planned add to a level of competence over the number of skills and abilities.

The message is if you have implemented systematic sewing operator training and are achieving results then make even better use of your valuable trainers by targeting training regimes to your products and circumstances. Do not take the "one size fits all" option but grow your training expertise.

Recording Operator's Work Start/Stop Time and Its Importance

Do you know when your stitching operators start stitching garments in the morning and when they stop working on the sewing machines before leaving the shop floor in the evening?

It seems a very difficult task recording all operator's actual work start and work stop time.

Is it practically possible capturing all operators exact work start time and work stop time?

Yes, this is possible with the help of technology solutions. With the manual aid, this much in-depth data capturing is not possible.

Currently, most of the garment units have CCTV on every corner of the factory and shop floor. A dedicated department or the factory owner (in small factories) keeps eye on the floor whether operators are working or not, what the operators are doing. CCTVs do some work on stopping operator doing unethical practice on the floor. But from the CCTV recording, one cannot create operators work start/stop time report.

Payroll clock-in (when employees enter the factory and when they leave the factory in the evening) data is not enough these days to improve the shop floor performance. As you do not know how many hours operators attended the work from the payroll clock-in/clock-out report.

Secondly, it is not possible for a supervisor policing operators everyday morning, whether operators started sewing garments after reaching to their workstation. Having a technology solution, a factory can keep eye on exactly when an operator is starting his/her sewing machine and starts sewing the first piece in a day.

Third, when the last piece is stitched, that time can be also captured. If one can record such data, you can show it to your operators. Then operators would not get chance to waste their productive time.

Why should one have accurate data for work start and work stop time?

It is true that in morning, operators need some time to warm up and catch the real stitching speed. If nobody is watching them when they are starting their work, they will take a longer time in starting stitching.

That is why the first-hour production quantity and produced SAH normally remains less compared to following hours.

In an assembly line, many operators will not have work to sew - they wait for getting work. If their first bundle start time getting noticed by managers/supervisors, they will pull work and start their machine.

You can reduce operator idle time by having and analyzing idle time. This, in turn, improves resource utilization and improve floor productivity.

Operator's non-standard efficiency and real efficiency calculation will be easier when you have clock-in and clock-out data. Work start and stop time is also called clock-in and clock-out time.

Technology for capturing operator's work start time/stop time data

Following technologies can help you capturing operator actual clock-in and clock-out time.

- Real-time shop floor control system can capture such timing
- A microchip embedded in the sewing machine can capture such data
- A programmable sewing machine that can record sewing machine running time (clock-in and calendar date).

Operator clock-in and clock-out report

Here, I will show you one basic report of sewing operators' clock-in and clock-out report, which is made from operators' login to their workstation.

Date	2-Feb-18	EMPLOYEE CLOCK-IN / CLOCK-OUT REPORT		
Line	Employee Code	Employee Name	Clocked-in Time	Clocked-out Time
Line-1	10001	PRAVEEN DAS	09:33	18:00
	10002	RANI DEVI	10:25	18:05
	10003	RAKESH KUMAR	09:30	13:31
	10004	REKHA DEVI	09:32	18:00
	10005	MANOJ MANDAL	10:04	10:56
	10006	GEETA MAJUNDAR	09:33	18:01
	10007	JUGENDER KUMAR	09:34	18:00
	10008	JITENDRA KUMAR	09:31	18:00
	10009	RAJ RANI	09:32	18:00
Line-2	10010	SEEMA GUPTA	09:31	17:59

Figure: A Sample Report: Operator's Clock-in/Clock-out Report

I had prepared this report for one garment manufacturing unit. Operator clocked-in and clocked-out to/from the workstation data. These data are taken from shop floor control system. From this report, a line supervisor, exactly knows which operator has started working at what time and till when they were on their workstation.

You can see the total working hour of an operator in a day. Every operator will not be at workstation for all 8 hours.

If you have database of operators' start/stop time, you can prepare many other report related to this. From the database, one can further prepare a report with exactly when the first garment stitching was started by the operators.

Should We Include Helpers in Line Efficiency Calculation?

Question: *I have read the line efficiency formula in your blog, but I want to know that apart from operators, the helpers who are working in that line are not being included in calculating the line efficiency. As they are part of that line, without them for sure operators cannot achieved what they can. So, it would be very helpful if you rectify my confusion.*

There is no standard rule whether to add the helpers' working hours in line efficiency calculation or not. It is up to engineers, how they present the efficiency data.

I have answered this question with explanation. Go through my point of views and decide whether you agree or disagree.

As discussed earlier, the following formula is used for calculating a line efficiency.

> Line Efficiency% = (Total minutes produced by the line / Total minutes worked by labors working in the line) **x** 100

You can see, for calculating line efficiency, we use two data-
- Minutes produced (You can say standard hours produced)
- Minutes worked (Total minutes worked by the employees)

Here the produced minutes is calculated as
Produced minutes= (Style SAM **x** Number of units produced)

When helpers are present and working in the line, they have working hours. But they do not produce any minutes (or produced hours) as most cases helpers are not given defined operations. Also, the kind of work they do in a sewing line, the SAM is not allocated for their helping work. Helpers are mostly involved in works such as line-loading task, moving bundle from one place to another, collecting and counting stitched garments. This task is considered as unmeasured work.

All helpers do not work on the job that can be measured.

They do not have the contribution in total minutes produced by the line (theoretically). If you consider helpers in the efficiency calculation, you are not balancing hours produced and hours worked. You do not get true line efficiency data. Although practically they are involved in the production and assist sewing operators directly or indirectly.

If you allocate standard minutes (SAM) for the task done by a helper, then you must include that helper in efficiency calculation. Like pressman, marker man's time is included in the style SAM and they are included in line efficiency.

You also need to understand the kind of task helpers are doing on the production line.

I know many top-tier garment manufacturing companies in India and Sri Lanka who do not include helpers' working hours in line efficiency calculation. You can say helpers work is not measured.

I hope my answer has cleared your doubt on whether to include helpers or not in calculating efficiency of a production line.

There is another question related to this this question. I have explained it here.

Why Indirect Manpower is not Included in the Efficiency Report?

Question: *Why we only add direct manpower in the efficiency report? Can't we add a quality person, input man, line leader, line supervisor i.e., indirect manpower? ... A question from an OCS reader.*

When you are talking about the efficiency report, it includes line efficiency and required information for calculating the line efficiency. Though you can keep a record of supporting manpower in the efficiency report without including their attended hours in efficiency calculation.

Efficiency is calculated as Hours produced against Hours worked. Or total minutes produced against total minutes worked (attended minutes) by the direct employees.

To get minutes produced from each employee, standard time need to be assigned to all jobs that are performed by direct and indirect manpower working in a line. But in practice, we cannot assign SAM for line supervisor's job and their job is not included in the style (operation bulletin). Therefore, we are not getting any produced hours from line supervisors and line leaders. That is the reason we do not include line supervisor, machine maintenance and line leaders in efficiency calculation.

In case a quality person checks garment and you have assigned SAM for quality checking task, you need include quality checkers and their attended hours in efficiency calculation. If checking job is not included in the OB, their work should not be considered in efficiency calculation.

Direct manpower: Direct manpower are those do the work and involved in making the garment. Sewing operators, helpers, pressmen, quality checkers are direct manpower (labor). In garment costing, their cost is considered as a direct labor cost.

Indirect manpower: Indirect Manpower are those who do the supporting role to get the job done. Mostly those doing the supervision activities, line leaders, administration team, line feeders, and industrial engineers.

Who Should Count Operator's Production?

In the manufacturing sector, measuring production of the line, or recording operators' production is a common practice. So, in the garment manufacturing, we capture daily production and keep the production record. My question is who should count the operator's production?

In many garment factories, workers count it and write their production quantity on a paper, (in one factory I had observed operators carry a notebook to record their daily production). Their notes (production counting) help them in measuring their work and their earning. Their notes help supervisors to check production target and actual production, in line balancing and production motoring. No doubt, it is a good practice.

Let us look at this practice from the engineering point of view. Where operators count their production after completing each bundle (in some operations after completing each piece), they waste their productive time. They are lowering their potential performance and daily earning. Which in turn reduces the line productivity and factory productivity. Reduction of profits and factory capacity. Operators are employed in making garments. Some of them sew the garment, other does marking job or pressing job. They are not supposed to count their hourly production or daily production.

Imagine how many hours will be lost if all operators in a floor count the number of pieces after completing bundle (or after each piece they complete) and write it in one paper throughout the day.

What will happen if the factory allows operators to concentrate on the work and forget about noting their production? Instead, the factory can hire a helper (or as many as required) for counting the hourly production of each operator.

An alternative way, a factory can install real-time production tracking system (either RFID technology or Barcode technology) for capturing production data of individual operators as well as line's production without wasting operators' standard time.

Data capturing should not be done with the cost of operators earning. Think on it!

Review Yearly Performance by using 5 Trend Charts

In the new year, many of you start new improvement projects for the betterment of the work and factory performance. Before you start a new improvement project, and become busy with the new assignments, it would be a good idea summarizing last year's data and review last year's performance of the factory.

It is time to look back and review the last year's performance. How much you have achieved in the previous year. What all KPIs are not meet your goal. In which all KPIs you have beaten the last year's benchmark performance. You can analyze the available data as many as you want.

A factory must review these 5 KPIs for the previous year. The last year's factory performance trends would help you in planning this year's goal. Let us prepare the last 12 months monthly performance trend and last 5 years yearly performance trend on the below KPIs

1. Line Efficiency Trends

Calculate monthly line efficiency and draw a trend chart on a spreadsheet with 12 months of efficiency data. Line efficiency is a common performance measure and I assume you calculate line efficiency, and you have the monthly efficiency data. If not, you can calculate it now. Read the earlier article on efficiency calculation method, to learn the line efficiency calculation method. Here is an example efficiency trend chart.

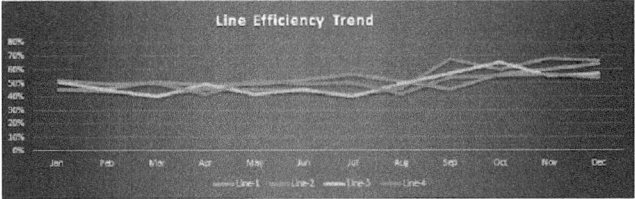

2. Production Volume and Shipped quantity

You earn revenue when you ship garments. More you ship, more you earn. Review the monthly production volume and shipped quantity.

Prepare a trend chart month by month. Here is an example of 100 machines factory that makes knits tops, t-shirts.

3. Cost per garment

To earn more profit, it is necessary to reduce garment production cost. With the continuous increase in wages, consumables, and machine maintenance cost, how the production cost changing month by month. Is your production cost per garment is meeting the buyer's target cost?

Assuming your factory is making a similar product range year after year. If not, still measuring cost per garment and analyzing its trend is a good way of understanding the factory's performance growth.

4. Cost per SAM (production cost per standard minute)

Cost per SAM is the labor cost per minute. Calculate actual labor cost per standard minute. Cost per SAM may increase due to many reasons. It may be increased due to the rise in minimum labor wages. Cost per SAM may higher in case line performance goes down due to some other reasons.

5. Quality performance (Percentage defective)

With productivity growth, you need to maintain the product quality standard. So, do not miss analyzing the quality performance data. For this calculate monthly percentage defective and draw a trend chart. You can also include the DHU report in the quality performance chart.

If you are new to measuring key performance indicators (KPI) of your factory, you can refer to my eBook – *"Garment Maker's KPI - Why measure and how to measure."*

Conclusion

Once you are ready with the above 5 performance trend charts, go through the charts. Charts will tell you what you have achieved in the last year. You can extend the list of KPIs for review. Factory management must consider measuring top KPIs and reviewing data.

It may be time to meet with your team and plan again to set-up a new performance benchmark for the current year.

Another objective of reviewing yearly performance result, you set-up a new benchmark for each KPI. Employee performance and departmental performance can be assessed through the yearly KPI review.

What is More Important- a Skilled Operator or a Sewing Machine?

When it comes for improving factory performance and improving workers performance in an apparel manufacturing unit, experts would say productivity can be improved either by improving operator's skill level or by purchasing an advanced automatic/semi-automatic sewing machine. That means you have two options. Here the question is which one you will prefer for your factory?

Simply, we will prefer the one that is more important to us and cost-effective.

An advanced machine could be -

- equipped with automatic thread cutter
- automatic sewing workstation for specific sewing operation that reduces manpower requirement

Here are a few thoughts for comparing these two.

- Purchasing new machines is a one-time investment plus maintenance cost. But when we hire an employee, we need to pay her every month.
- We can buy an advanced sewing machine easily. To operate the machine, we need at least one operator.
- An operator can be highly skilled in one type of machines and low skilled in another sewing machine.
- An operator can operate all types of sewing machines after getting training but by using one single machine, you cannot make the garment.
- By improving operators' skill, one can improve labor productivity with the existing resources (manpower and machines)
- By adding more machines (similar machines) production can be increased but we can't say that labor productivity will be increased.
- An advanced machine can help in increasing production for some specific operations. But remember most of the time special machines are not fully utilized due to style difference and order volume.

- Where there is a shortage of workers (sewing operators) and workers are costlier, advanced machines can be the right choice for running a manufacturing unit.

Skilled operators and sewing machines both are important resources for a garment manufacturing company. The importance is how we utilize the resources.

Change Factory Floor Layout and Give it a New Look

How often a garment factory should re-design their floor layout?

Many garment industry folks used to say that when consultants come to a factory, they suggest a new way of doing the work for achieving a better productivity. This may be true.

I am writing this post after observing one factory for the last 10 years. In the last 10 years, the factory has rearranged sewing floor, cutting floor and other section many times. I think changing the floor layout time to time is good for many reasons.

Whatever way you design the floor layout, it has some benefits. I would say, it is not only for productivity improvement. I can relate this with the good feelings of workers. When feel good is improved, production increases automatically. The floor becomes livelier when you rearrange the production floor.

How to make changes?

You can take the expert's suggestion for layout change (workplace re-engineering) of the production floor. Here, I have listed a few pointers that can be useful for changing floor layout without expert's guide (these are my suggestions for doing things differently, not instruction).

- Moving small work section from one floor to another floor. Or from one corner to another corner on the same floor.
- Change directions of the material flow or operator seating direction.
- Change cuttings loading place. It is not necessary that you need to keep one table at the start of the line. You can keep the loading table in one corner of the floor.
- Use trolleys for carrying bundles.
- Make it a straight line if you have different line layout.
- Make it a U-shaped line if you are following a straight line for couple of years.
- Set-up inline finishing and inline washing section.
- Make garment parts in a separate section (if you have preparator oprations)

- Make parts in the main production line (if you do it in a separate section).
- Relocate the fabric checking tables and fabric racks in the fabric store.
- Upgrade the office area. Why not make the office space like an IT office?
- Remove cabins for HOD and managers.
- If the full line is not utilized for a style, set-up the line near the output area.

Benefits of changing production floor layout

I can visualize the following benefits.
- improves labor productivity
- the floor becomes neat and clean as all the clutters and unwanted items lying on the floor will be removed.
- the floor becomes livelier. Improves lighting after cleaning and improves airflow and air circulation.
- All the factory staffs and workers get something to discuss. Many employees get motivated after doing the tough job of changing the floor layout
- improves teamwork and relationship between workers.
- Continuous improvement happened

Did you change the floor layout in your factory after the first set-up?

Try this and do it regularly. You will see more energy flowing inside the production floor and in the factory.

You might be thinking of time loss for redesigning line layout. Choose the season when you have less work on the shop floor. The cost incurred in doing changes - consider this a maintenance activity.

Whitewashing the walls and ceiling once in a year is not a bad idea.

Have you recently changed the production floor layout in a garment factory?

Real-time System for Production Tracking System in Apparel Manufacturing

For real-time shop floor production tracking and production monitoring in apparel manufacturing, RFID technology based system is one of the commonly used systems. I will be sharing about the RFID based real-time system used in a garment factory for real-time production tracking and shop floor monitoring. RFID stands for Radio Frequency Identification.

I do not need to mention that we all like faster product delivery and quick response to the market. We use many mobile applications in our smart mobiles for managing our personal life. When the apparel industry going for smart manufacturing, machine learning, using IoT, AI tools, apparel manufacturers feel the importance of data for managing their production floor - Why we can't have real-time production tracking system for the production floor to manage the production in a better way?

When you do manual production tracking, you count garment production hourly or bi-hourly. You count production from the end-of-line output. You count sewing operators and helpers in a line. You update production quantity on the Hourly Production board. You get production status of the whole factory after one day. That is too late for decision making and fixing production problems. What if you get the production update after each garment is stitched and completed in each workstation?

Let me explain what we mean by production tracking in a garment factory. Production tracking means more than only counting pieces at the output operation and operator head counting. Getting production counting from each workstation (operation-wise as well as operator wise production data), knowing how operators spend the given time when they are present in on the shop floor.

RFID based production tracking system is an old technology, still, it has the potential for garment factories for automizing production tracking and reporting.

Purpose of the Real-time system for apparel

The primary objective of an RFID based real-time production tracking system is capturing real-time data and viewing the report on real-time.

Improving process visibility and transparency. Accuracy in production data.

Data capturing and getting data automatically through scanning RFID tags from each sewing workstation. Once you have captured data from the desired activity in the pre-defined parameters, you can play with the data and build reports and KPIs as you need.

Benefits of the Real-time Systems in production tracking

When you have a real-time production tracking system, you will have a database. You have the flexibility of making reports in many ways as you want. Factories those have strong IT team and having knowledge of programming and making a report from SQL database, can easily make useful reports for the management and factory. Customized production reports can be made in Crystal report, Excel-based reports using SQL query.

With this system, data recorder and line supervisor cannot give you inaccurate production data. As each operator gets RFID tags for the garments she stitched.

Manual data collection and report making can be eliminated by installing a real-time production tracking system. All production data, lost time data and quality data will be captured from the operator's device and the system will process those data and report will be made automatically in the pre-fined report templates.

Improved visibility and transparency. You can view production data from a remotely having internet connection. Managers can review data for all sections and lines. They can compare performance between sewing lines. Can track order status, WIP.

Some of the common reports factories views include
- Hourly production report
- Line wise employee attendance and absenteeism report
- Previous day's employee-wise production, operation-wise production report
- Today's line efficiency and previous day's line efficiency report
- WIP report
- Individual operator's efficiency and skill history report

With the real-time production tracking system, you can introduce an incentive scheme for sewing operators. You know incentive scheme is the main driving force for productivity improvement in the shop floor. For designing an incentive scheme, you need production and efficiency for individual employees. With the real-time production tracking system, you will have all the basic data you need for incentive scheme and calculating employees bonus amount. Operator incentive can be designed in various ways – like individual performance, group performance, a combination of group and individual performance.

You can eliminate the bar-code system on the production floor in case you are already having such a system.

Your operators do not need to write production quantity in a piece of paper.

You can integrate other modules in the production tracking systems - like real-time quality control system, machine maintenance module.

RFID tags are reused until the tag get damaged. Even one tag can be used for more than 10 years.

A time-saving tool for industrial engineers. In a garment factory, IEs spend a lot of time in daily report preparation and data analysis. Data preparation can be eliminated by introducing a real-time system.

Improving productivity – All application and software solution provider commit that it will increase the factory's productivity. But how? A system can only provide data – you need to take action to get the benefit of a system.

Lost-time data – by capturing lost time data, you can find major reasons for efficiency losses in production. After getting lost-time data, if you act and reduce lost-time, your productivity will improve.

Limitations and challenges of a real-time system in apparel manufacturing

Each factory is unique, and different garment factories view daily production reports in different formats. The inbuilt reports in a real-time production tracking system are limited. In case the factory wants additional report matching with their existing excel reports, suppliers can develop the customized report with/without additional charges.

Involvement of the line supervisors, line leaders and sewing operators is essential for a successful implementation. Sometimes questions come – why operators should spend time on scanning RFID and follow the new system if they are not getting any benefits from this system (in the salaried environment).

I have seen real-time production tracking systems is used as standalone software and not integrated with the ERP and payroll systems. Engineers become the lead for the system installation and maintaining the system. When an IE leaves the factory, factory face issues in maintaining the system.

During the implementations, factories develop many custom reports. But after a few months, they used to view and track only a few reports.

Most of the factories I had worked with do not have a database team (I mean, IT team having knowledge of SQL DB and report preparation from SQL database) for developing reports internally. In such a situation, factories cannot utilize the data they captured.

Conclusion

I understand it is not possible to understand and visualize the whole process of a real-time system and its use in a garment unit by reading this short article. I have explained only major activities with brief notes and without any example. My objective of sharing this piece of article is to make you aware of how the real-time system works in an apparel manufacturing unit.

The technical things and actual working method can be learned when you install such system in your factory. There are a few companies who provide real-time production tracking solution, most of them use RFID technology for data capturing. Working method and data preparation may vary from one product to another.

For tracking the real-time shop floor production data, RFID systems are good options for the garment factories. Benefits of this system are enormous if you can fully utilize the product.

Software to Establish Garment Standard Time

In the apparel industry, knowing the accurate standard minute of a product is very important for many reasons. There are mechanisms for determining product SAM other than software, like Work Sampling and Time Study. But time and motion study-based calculations took a longer time to establish the standard time for a garment. Secondly, there is always a question of getting the correct SAM established through Time Study and operator performance rating method.

To get rid of these issues leading garment manufacturers are adopting SAM calculation Software. With quick Standard Minute establishing, most of these software products help in calculating direct labor cost and CMT price; setting up standard sewing methods and improve production methods.

In this post, I have shared leading software products those are available for establishing standard time. All the following products are developed from Methods Time Measurement (MTM) and known as PMTS database. (the following list in not in particular order)

1. General Sewing Data (GSD) / GSD Cost

GSD is popular and globally accepted product for work measurement and setting up standard time for garment sewing operations (machine operations and manual operations). GSD is developed by GSD Limited, UK. At present, GSD is owned by Coats.

2. Standard Sewing Data (SSD)

SSD is now named as TimeSSD. SSD is developed by AJ Consultants. Standard Sewing Data (SSD) system suits to all sewing rooms from light to heavy – from underwear to upholstery. With the universal Standard Work Data extension, the system suits to all manual and tool-assisted operations in a garment factory.

timeSSD standard data and methods development program is for creating accurate time standards fast – and in advance – for labor costing, production planning and work loading and incentive and piece-rate systems and for continuous methods development and efficient job training. SSD is fast and easy to learn and to use.

3. SewEasy® Quick Garment Sewing Data

SewEasy is developed by Sri Lanka based company SewEasy. SewEasy® Quick Garment Sewing Data system is used for estimating the Standard Minute Value (SMV, SAM), a measure of labor content of garments, home furnishing, lingerie, leather goods and all sewn products.

4. Pro- SMV

Pro-SMV is the tool to establish Standard Time and Best Methods of sewing operations. This product is developed by Methods Workshop Apparel Consultancy. Now, this product is based in India.

5. MODSEW (Standard Data Apparel Engineering System)

MODSEW is a computerized adaptation of MODAPTS, or MODular Arrangement of Predetermined Time Standards. MODAPTS is used for calculating reliable production standards, improving an organization's productivity, analyzing departmental efficiency, and improving employee relations.

If you need to know more about these software, you can get in touch with the solution providers.

About Prasanta Sarkar

Prasanta Sarkar is a garment industry professional, an author, and an employee in a software company.

He started writing one decade ago while working in an apparel management consulting firm, in India. He joined a garment manufacturing export house after completing his post-graduation in fashion technology course from National Institute of Fashion Technology, New Delhi. He has a B.Tech. degree in Textile Technology from Calcutta University.

In the year 2011, he had launched the blog Online Clothing Study (www.onlineclothingstudy.com) and since then he is sharing his learning through his notes and articles. He helped learners to become knowledgeable and informed in the field of apparel manufacturing and apparel supply chain.

He has written and published 5 books and this one is the sixth book in his credit. Many garment businesses and newcomers got benefitted from his blogs and guides.

www.ingramcontent.com/pod-product-compliance
Lightning Source LLC
Chambersburg PA
CBHW072026230526
45466CB00020B/933